I0434423

SCIENCE THAT TOUCHES LIFE

THE DEEPER
MEANING OF MATHEMATICS

MUSA YILDIRIM

The Science Touching Life

The Deep Meaning of Mathematics

MUSA YILDIRIM

Shape and Graphic Design: Tuncay Canpolat

ISBN: 979-888-459-71-74

Resume

Musa YILDIRIM completed his primary education at Bahçelievler Secondary School in 1992 and started his secondary education at Cumhuriyet Science High School in the same year. After his secondary education, he started his higher education at Istanbul Technical University, Department of Aeronautical Engineering in 1995 and graduated from this department in 2000.

Following his military service, he joined the Air Force Command as an aircraft engineer officer. During his years of service in the Armed Forces, he worked as a system engineer and project manager in many aviation projects for F-16, C-130, C-160, KC-135, SF-260D and T-41D aircraft and UH-1H and AS-532 Cougar helicopters. In 2008, he completed his master's degree in mechanical engineering at Erciyes University. In 2020, he retired from the Air Force Command and in the same year, he started working as a "Senior Expert Design Engineer" at TAI Aerospace Industries.

In 2016, Musa YILDIRIM gave lectures at the Aviation and Space Workshop organized by TÜBİTAK and at the aviation seminar organized by the Middle East Technical University on the benefits of the National Combat Aircraft Project for the defense industry.

He also made presentations at the Presidency of Defense Industry and TAI Academy within the scope of the book study *Science Touching Life - The Deep Meaning of Mathematics.*

Musa YILDIRIM, as a person who believes that the only way for our country to be able to compete with the states of the world is through a good mathematics education, continues to talk about this subject in every environment.

Acknowledgements

This book has emerged at the end of a challenging seven-year process. I would like to thank my dear wife Gülender YILDIRIM and my children who never left me alone and never withheld their support, my dear friend Tuncay CANPOLAT who did the graphic design works of the book and all my friends and colleagues who supported me in writing this book.

TABLE OF CONTENTS

This book,
my children whom I love more than life itself;
I dedicate it to Azra, Mekser Berk and Elif Irem.
I hope they will do useful and beautiful work for my beautiful country
and wish

Getting started...

When asked the question, "Why is mathematics taught?", those who have licked some ink begin to explain that it is taught to raise individuals with analytical thinking. When you see the mathematical equations, we use in elementary school to solve problems such as pool, age, speed, interest, labor, etc., you realize that mathematics actually has a counterpart in this life, and when you solve them, you start to say, "This is why we learn mathematics!".

Of course, solving simple problems such as pool, age, speed, interest, labor, etc. is really enjoyable and it is like a short movie screening that makes you feel like you are learning mathematics to solve them. When you solve them, you get a little bit of the feeling that *"what you have learned has a purpose"*. Unfortunately, this feeling of satisfaction is enough for a lifetime and - paradoxically - you stop questioning the education system because of it. You continue to memorize the subjects by solving simple problem questions and without questioning the lectures too much.

In fact, this is the trailer of the main movie you are watching, but you cannot watch the trailer and act as if you have seen the whole movie. Of course, the trailer gives you serious clues about the movie. But you can't get the pleasure, the taste you get from the movie from the trailer.

As you solve simple problems, your analytical thinking gradually develops and matures. Entering the world of real problems from here is a great start in terms of the structure of education. However, without transferring mathematics to real problems in this life, you cannot survive in this life by solving only such simple problems with what you have learned. If you don't use this knowledge to find the real problems hidden in the depths of life, the mathematics you have learned will not tell you much.

The question, "Teacher, where will this help us?" has no meaning on its own. Because you can live and continue your life without knowing math. Aren't there people who live with almost no need for math? Of course, there are, just look at underdeveloped countries to understand this. What is actually being asked here is how you can use the mathematics you have learned to build automobiles, airplanes, engines, bridges, dams, combine harvesters, skyscrapers, subways, telephones, computers, satellites, and how you can use mathematics to build technological products such as these.

We are always told that there is serious mathematics behind all advanced systems, but we often fail to see it. The curriculum is always blamed for this, but the curriculum is not to blame for this. Wherever you go in the world, you see more or less the same curriculum. Almost everywhere education is done with this curriculum. Isn't the main reason why the Western world can build space shuttles with the same curriculum and we can't do it is because we learn these things superficially and memorize them?

We are told that mathematics is the language of science, but not much information is shared with us about how this happens in our educational life. Since we cannot establish a serious connection between disciplines, we continue our lives in an education system that constantly tells us but never shows us the connection between the mathematics we learn with memorized cliché expressions and life. This book was written for you to see and perceive what mathematics actually wants to tell you in a unique way by going beyond the memorized teachings.

Everyone learns mathematics, but I can say that the number of people who know why they learn it is very, very small. If we learn the background of mathematics, that is, its purpose and where it wants to take us, I can say that many things will change for my country. Because countries that know the background of mathematics very well rule the world. So much so that we see that they have very good mathematical knowledge everywhere, from their movies to their music, from their paintings to their architectural structures, from their financial systems to their city plans. You cannot make original products by copying what they do. Where there is copying, there is no mathematics. This is not how you learn mathematics anyway.

This book is the product of an intellectual period that started with the idea of finding a solution to the problem of not being able to bring mathematics into life. In the book, I tried to explain that mathematics develops by touching life, that all the rules and principles emerged for this main purpose, and that they are built on the four basic operations of *multiplication, division, addition* and *subtraction.* I wanted you to see the background of the basic teachings of mathematics. I tried to reveal the subjects that are learned by everyone but whose back side is not known or seen much, and to underline the subjects that are constantly explained everywhere but whose back side is not emphasized much, and to make them transferable into life.

When you look carefully at the side of mathematics that touches life, you immediately realize that mathematics is a universal language and that mathematical rules are not formed in vain. We owe a lot to great mathematicians such as Newton, Einstein, Leibniz, Euler, Gauss, Tesla and Maxwell who helped the language of mathematics to develop and enrich. I don't think I need to tell you that we are what we are today thanks to them. In this book, I have tried to make the facts revealed by these great scientists visible and understandable in all their nakedness. In fact, I wrote this book to seek an answer to the question "Why do we really learn mathematics?".

Since I see that the lack of such a book is a serious deficiency for my country, I have tried to explain how the analytical point of view emerged and how it should be by writing what I have been thinking for a long time. I say what I think, but of course I was also educated with these teachings. Without going beyond the subject, I just tried to make the existing ones more visible. I know that I have shortcomings. Of course, I cannot fit a universal language such as mathematics into such a small volume book. But people who read this book will see that once they get the main idea of this book, they will be able to penetrate into all subjects of mathematics very easily.

Readers who get the gist of the book will be able to grasp and interpret the topics covered much faster, whether they are in basic sciences, engineering or social sciences. When you get the spirit of mathematics, you will see that the topics covered are not very difficult. In fact, you will realize and regret that you cannot look at the world and the universe properly because you do not really know mathematics very well.

When you actually hear and feel the things described in this book, you will realize that the science that touches life is not so far away, that it is everywhere when you look around you a little more carefully, and you will be very happy. You will look at the path a piece of paper follows as it falls to the ground, the change in the number of push-ups you do when you increase your weight, the physics of recording sound or a picture on a metal, the air power required for an airplane to fly, and you will embrace mathematics with all your might to understand and interpret them. Then the mathematics you learn will touch your life and you will start to enjoy it more.

You will look at the formulas you have learned in science and social sciences in a completely different way and you will understand and interpret what they mean much better. You will taste the feeling of "How do I make a mathematical model of a problem, how do I define its functional relationships, how do I transfer it to a coordinate system, in short, how do I solve it". This book was written to trigger this. You will understand what it means to know the world and the universe under the guidance of mathematics. That's when you will say "I am glad there is mathematics, I am glad I learned mathematics!".

You will be able to ask and answer original questions on scientific topics without looking at the answers to the questions. You will not have to look at the answer key in foreign sources to see if your answer is correct. You will overcome your fears about mathematics, write your own answer key thanks to this book and say "Mathematics is what mathematics says it is!".

Importance of Mathematics

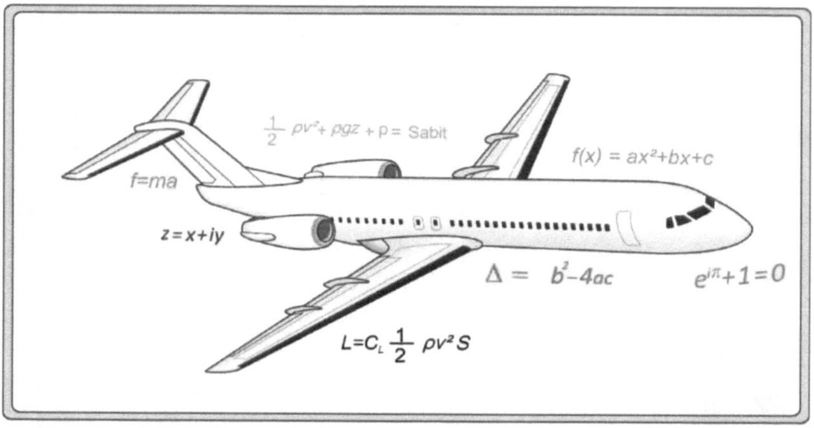

Why is math so important?

Mathematics is the language of science. By using this language, we understand and decipher life. Without mathematics, we cannot solve the codes of this life with the words and sentences we use only in daily life. We can say that mathematics has entered our lives since the first human being came into the world. Of course, there was mathematics before humans and we try to explain everything we can observe with our mathematical knowledge. We constantly observe this knowledge as we live. From the shapes of celestial bodies to their movements, from the growth of plants to the movements of air and water, many events tell us this in their own mathematical language.

In times when there was no money, we can consider the exchanges in trade as simple mathematical operations, can't we? Wasn't the value of each commodity, what it was worth, determined through exchange? Didn't the concept of equality, which would later become one of the most important symbols of mathematics, play a critical role here? After the invention of money, the value of a good was determined according to the weight and shape of a metal. Gold, for example, has been used as a medium of exchange for thousands of years and continues to be used. In fact, mathematics has tried many ways and methods to show itself to people, but unfortunately, we had to wait for many years to see and find the mathematics in life.

For thousands of years before numbers, we used words like "small, big, wide, narrow, long, short, little, a lot, a lot, quite a lot, low, high, not bad, okay" to try to make sense of expressions and magnitudes in life. But with the introduction of numbers, these expressions and magnitudes gained their true identity.

Thanks to operations such as addition, subtraction, multiplication and division that came into our lives after numbers, we started to play with numbers like balls, and we started to overcome more complex expressions by converting them into each other. Now, questions such as "How much more? How many times this?" had an answer for us.

But the developments in mathematics in the last 300 years have given us results that are perhaps a thousand times more meaningful and valuable than the developments that took place in the previous tens of thousands, perhaps millions of years. The number of problems solved with mathematics learned until the 18th century is nothing compared to the number of problems solved after this century.

When you look back into the past, human beings have made a wide variety of tools, from the wheel to the spear, from the arrow to the bow and sword. Unfortunately, it took thousands of years to make them. When you see the tools and machines made in the last 300 years, it seems as if the people of the past did nothing, doesn't it?

Humans only found iron, fire, paper and writing thousands of years after they came into the world. We now know that the first people lived in caves or forests. How long do you think it took for human beings to build a house and settle down? If mankind has been on this earth for ten thousand generations, we can say that perhaps 90% of this time mankind has been living in caves.

Figure 1.1: Innovations in our lives thanks to mathematics.

The wheel was invented 5000 years ago, but how long did it take to move from this invention to the car? The car was invented not even 150 years ago. Pay attention to the disproportionate difference between the years, and you will understand better what I mean. When you multiply such questions, every scientific achievement of the last 300 years tells us a truth.

What happened in 300 years that so many mathematical terms and expressions were produced, derived and entered our lives? We have already seen that it takes a few centuries for these terms and expressions to turn into airplanes, tanks, satellites and computers.

In fact, math was everywhere, but people had to wait thousands of years in vain to see it. What has happened in the last 300 years is that mathematics has entered into everything inside and outside matter.

This is what has happened in the last 300 years, that mathematics has touched life. People all over the world have seen this, and it has enriched mathematics and strengthened its connection with life.

Figure 1.2: The language of numbers changed everything.

You may not need much mathematics when making a bow, spear or arrow. When you look at the tools and structures built in the past, we can say that these tools or structures were built with very little mathematical knowledge. Yes, of course, you need to know mathematics to build the pyramids in Egypt or the Great Wall of China, but this mathematical knowledge can never come close to the mathematics used in the construction of an airplane, a submarine or a satellite.

We can explain the basic philosophy underlying the developments of the last 300 years as the introduction of mathematics into the scientific thought system, which took on a completely different form and gradually enriched it. Yes, mathematics has not been idle in this process and has constantly improved itself.

From logarithms to complex numbers, from functions to series, many mathematical operations and their notations enriched the mathematical language. We can easily say that the increase in the number of mathematical concepts in this period has a proportional relationship with the works produced in this period.

Of course, the enrichment in the language of mathematics was triggered by the needs of people in their research processes during this period. Instead of heating water and knowing that it boils by looking at its bubbling, much more meaningful pictures began to emerge by heating the water and seeing at what temperature it boiled and putting this into numbers and graphs. Expressions measured in this way and put into numbers and graphs began to tell us different things. We started to say sentences like "Water boils at 100 degrees Celsius!" The number of these sentences increased as the experiments were carried out to such an extent that we could no longer calculate them.

Just as we can distinguish people from each other by giving them names, similarly, we started to classify millions of events in our lives thanks to mathematics. Over time, we determined the boiling and freezing temperatures of other substances by using water as a reference. Heating or cooling matter are just two of the physical phenomena! With experimental studies, the era of giving value and meaning to natural phenomena has begun. It was very easy for humans to manage expressions in numbers, and this is what has happened in the last 300 years.

We said mathematics is the language of science... We started to express everything that was described in the past with expressions such as "less, more, big, small" with mathematical concepts. In the meantime, as the four operations and geometry combined with mathematics and entered our lives more and more, new concepts and new symbols began to be derived to solve the mystery of life.

Of course, with geometry, human beings have made this universe full of three-dimensional shapes more understandable. Thanks to geometry, the universe and, of course, all the events in this world can be put on paper. With the pictures we drew using triangles, rectangles, squares, parallelograms and circles, we were able to put this life on paper in a more meaningful way.

When you examine shapes in life, at first glance these shapes may seem very complex. But when you cut them up into neat shapes such as triangles and rectangles, you can easily see that more meaningful expressions are obtained. So, don't let the complexity of these shapes cloud your mind. You should sculpt these shapes in such a way that you should be able to see that smooth geometric shapes can be easily obtained from these shapes. When you approach the subject with this point of view, then you can better see that the complex structures in life can be easily overcome by using mathematics and geometry.

If you ask what the scientists of the Enlightenment Period have in common, look at physicists: They are all math and physics scholars... Look at chemists: they are all chemistry and math scholars... Look at biologists: they are all biology and math scholars... You cannot take mathematics out of any branch of science. In other words, you cannot become a physics, chemistry and biology scholar by removing mathematics. If mathematics did not permeate physics, chemistry and biology, we would never know what they are trying to tell us.

Examine the life stories of scientists such as Newton, Einstein, Maxwell, Gauss, Euler. You will see that you don't need to have a superior intellect to be like these people, they were just people who pursued curiosity and were able to integrate the mathematics they learned into the subjects they were curious about. Of course, these scientists are intelligent people, but in history you can see thousands of people with similar intelligence. But none of them found the truths that a Newton or an Einstein found. I will try to explain to you in this book that the reason for this is in the deep meaning of mathematics, in the details of the analytical point of view.

I would like to summarize the basic motto of this book as follows; if you know well how subjects such as integral, derivative, function, logarithm and matrix emerged, how their rules were shaped and of course why they were learned, you can place them into real problems and you can easily see how to solve problems that seem very difficult.

If you want to produce original scientific works, you need to know very well what the concepts that have been found and emerged in the past are for and where they are used. No expression, no formula came out of nowhere. If you do not understand the logic behind them, you will have difficulty in applying what you have learned. When you memorize what is taught in the lessons, you think you have learned the subject. When you cannot solve a real problem, you realize that what you have learned is nothing more than memorizing a few symbols, letters and rules.

Nor should it be concluded that "Let us learn the deep meaning of the existing mathematical methodology, that is enough for us!" The people who enriched the language of mathematics, of course, enriched the language of mathematics by using the existing methodology. You can make a paradigm shift and radically change the whole methodology of mathematics. In other words, you can walk through a different chain of mathematical relationships. If you change the mathematical methodology that is currently taught, the existing concepts will also change and you can even develop a different mathematical language. Of course, there may be concepts that will not change. In other words, concepts such as circle, circle, square, rectangle, triangle may remain the same, but you can produce and derive different symbols, expressions and functions from them. For example, you can call the circumference of a circle 365 degrees instead of 360 degrees, you can transform these degrees into each other with a different rule instead of arithmetic increase, you can build a new number system that will make complex numbers represented in two dimensions three-dimensional, you can switch to a different coordinate system that fundamentally changes the vertical coordinate system.

Figure 1.3: When you think about it, even your own body can tell you about mathematics.

I can summarize the main idea of this book as "Seeking an answer to the question of why we learn these things!". If you know why we learn these things, you will understand where they should be used. And when you use mathematics properly, you will contribute to uncovering the truths in life. Now, let's look for an answer to the question "Why do we learn the important mathematical concepts that are explained to us?".

Coordinate System

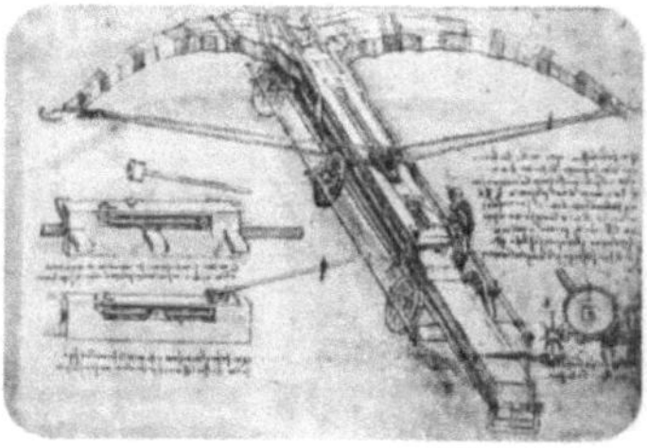

The Place of Geometry in Analytical Thought

and Birth of the Coordinate System

When we teach mathematics, we need geometry in so many places that I can't help thinking that many mathematical concepts emerged to analyze geometry. Whether it is for basic sciences such as physics, chemistry, biology or for different disciplines such as finance and engineering, you can see those basic mathematical concepts such as function, logarithm, derivative, integral, matrix have emerged and enriched over time in order to analyze geometric shapes in the coordinate system where all scientific subjects are conveyed.

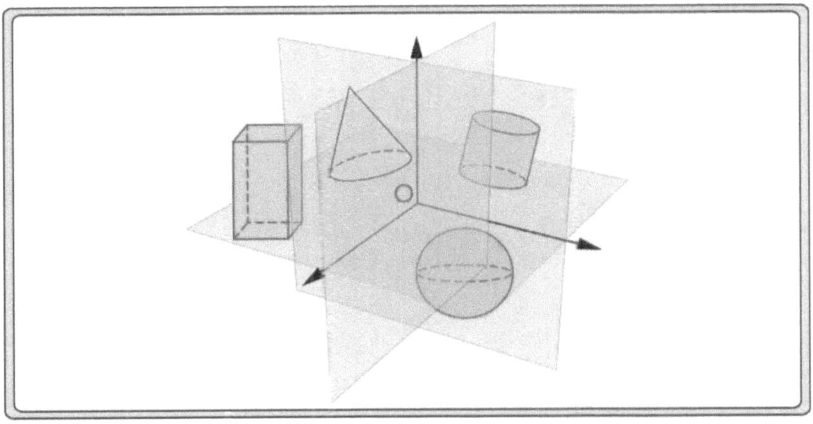

Figure 2.1: Geometries on a coordinate system speak volumes.

If you look carefully at your surroundings, you will see that there are actually three-dimensional geometric shapes everywhere. Of course, at first glance, you may not immediately see regular geometric shapes with definite width, length and height. When you look around you with concepts such as limit and rounding, which I will explain in detail later, you can easily say that we live in a world full of regular geometric shapes. Analytical thinking was born and developed in order to draw meaningful relationships and conclusions from these shapes.

One of the most basic teachings and rules of analytical thinking is to be able to visualize a problem while defining it. Your biggest helper to visualize your problem will of course be geometry. Geometry is the name of the science that tells us that in order to draw meaningful conclusions from this life, it cannot be done only with symbols, letters and numbers.

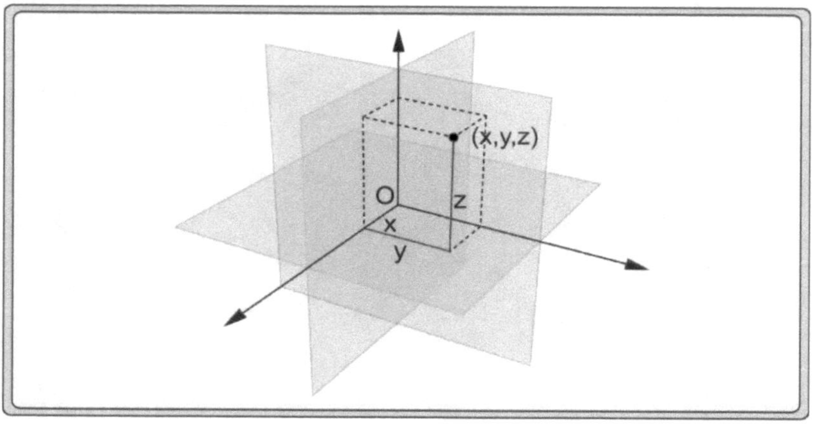

Figure 2.2: The coordinate system, where everything makes sense.

If you ask, "Where is the place where symbols, letters and numbers meet geometric shapes the most?" I will answer you directly as "Coordinate System". The coordinate system is the name of the place where the mathematical representation formed by two, sometimes three number lines cutting each other at right angles is staged.

The coordinate system, introduced to the world of science by Descartes, one of the architects of the enlightenment period in Europe, opened the door to a new world in the analysis of problems. By transferring a geometric object to the coordinate system, many definitions could be made and ideas could be obtained for solving a problem. The coordinate system is the name of the stage that offers many different solutions for your problem.

When you examine the life story of Descartes, you see that he was a great thinker and scientist who uttered the words "I think, therefore I am!". Descartes' emphasis on thinking as the raison d'être of human beings was a great revolution in the Renaissance period in order to free people from dogmatic memorization. In fact, this invention is as valuable as Euler's *functional notation* or Newton's *laws of motion.*

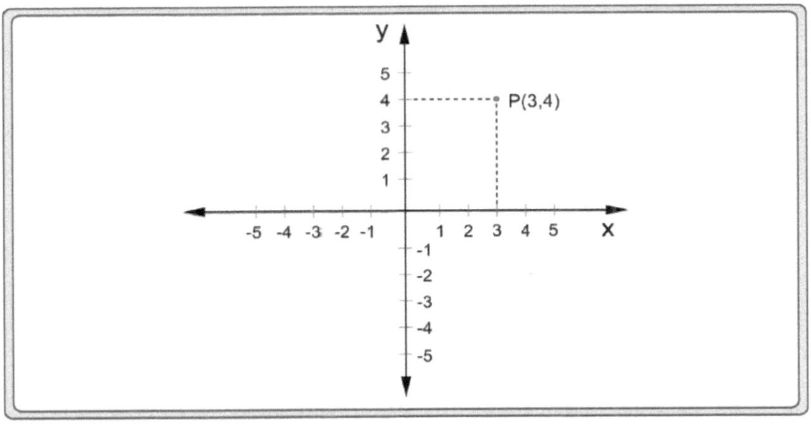

Figure 2.3: A simple coordinate system representation.

With the active use of the coordinate system, new concepts have been derived from mathematical functions over time and existing concepts have become more understandable. The combination of trigonometry and the coordinate system led to sine and cosine functions and made motion problems more understandable. In fact, you could even say, "Thanks to the coordinate system, complex numbers were invented." When you examine the history of complex numbers, you will immediately see that the coordinate system contributed greatly to the definition of these numbers.

Geometric shapes tell more than letters and numbers. By taking geometric shapes from the life we live, obtaining functions and drawing graphs, you see how problems that cannot be understood with numbers, letters and symbols become understandable over time. This alone is enough to explain the value of the coordinate system!

When you question how the coordinate system is taught in our current education system, you will come across information that there are areas of use in daily life such as solving the mystery of space, navigating in maritime, learning location information by satellite and drawing maps. If you don't know what a concept does and where it is used, bring the subject to space and no one will question the rest.

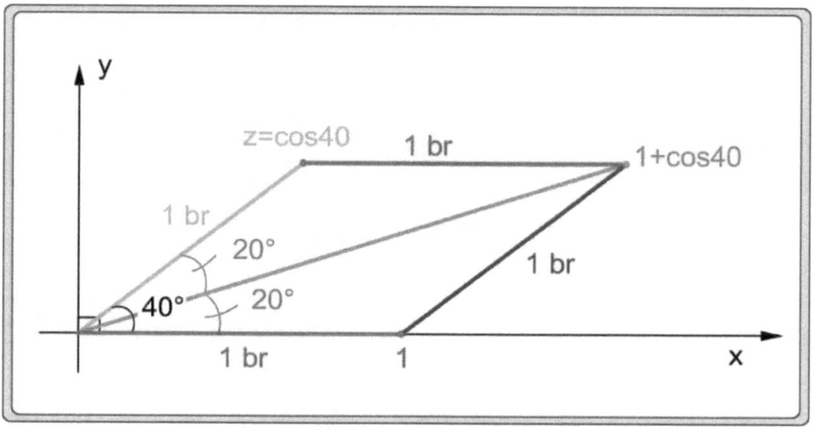

Figure 2.4: Everything is understood with lines on the coordinate system.

You can infer from this statement that if you don't want to deal with these things, you don't need the coordinate system at all. When I read such answers, unfortunately, I see that those who are explaining the subject do not understand the subject and I feel sad. The coordinate system is not only in the field of mathematics; it is the name of a scene that has entered the most detailed parts of physics, chemistry and biology. "It is used in space" does not necessarily mean "This subject is very important and must be learned". In fact, such answers prevent the subject from being learned for a universal purpose. To say that the coordinate system is used for a specific subject or for solving problems that people are not interested in is, in my opinion, not knowing what the coordinate system is for.

When mathematics is combined with geometry, it acquires a visual artistic identity. If you want to transfer anything you see or imagine to paper, the definition of geometry on the coordinate axis will make your job easier. Therefore, when you know why these concepts are produced and where they are used, you will not falter when you encounter any problem.

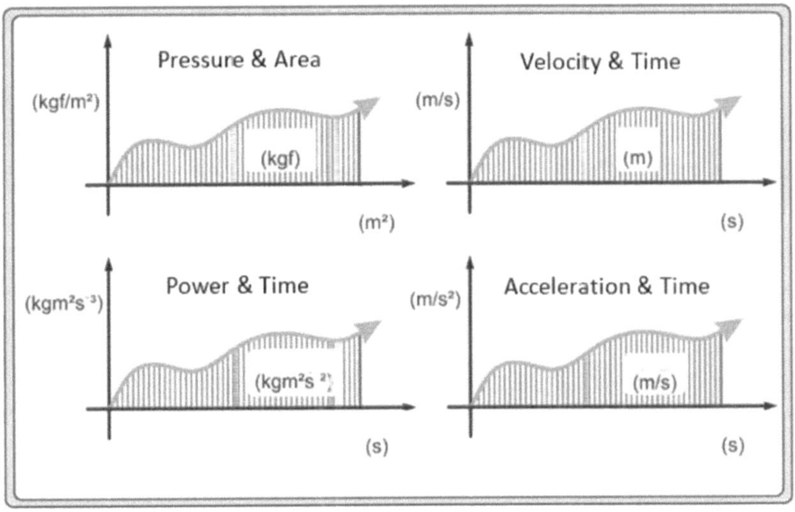

Figure 2.5: The coordinate system is the name of the place where formulas meet and formulas find their meaning.

When you plot a function in a coordinate system, that function may have been used in the definition of a problem, but when you take the integral or derivative of that function, you may have obtained another function. Even if the physical meaning is very different, the graphs you draw with functions always tell the same thing in the world of mathematics. You use the derivative and integral a lot when analyzing them. In short, whatever your physical problem is, mathematics does not deal with its units. It deals with the pictures, geometric shapes, their areas, volumes and edges that emerge when graphs are drawn, and extracts meaningful expressions from them.

Don't think that all functional relationships are found and finished! There are plenty of functional relationships waiting to be solved in this life, enough for everyone for thousands of years. If you find and extract them and transfer them to the coordinate system, only mathematics is enough for you to analyze the geometric shapes there.

We usually analyze the graphs we use in problems on coordinate axes defined by perpendicular intersecting lines. Of course, there have been efforts in the past to make graphs more meaningful with oblique intersections of lines For now, we can see and understand the picture more easily on perpendicular intersecting lines.

If you want to go to the next level in scientific thinking, intersect the coordinate axes obliquely and you get different graphs, or even have the coordinate axes intersect obliquely. You can try many coordinate systems like these. In the future, maybe we will solve problems with coordinate systems defined in a very different way. But for now, let's solve our problems using the perpendicular intersecting coordinate system.

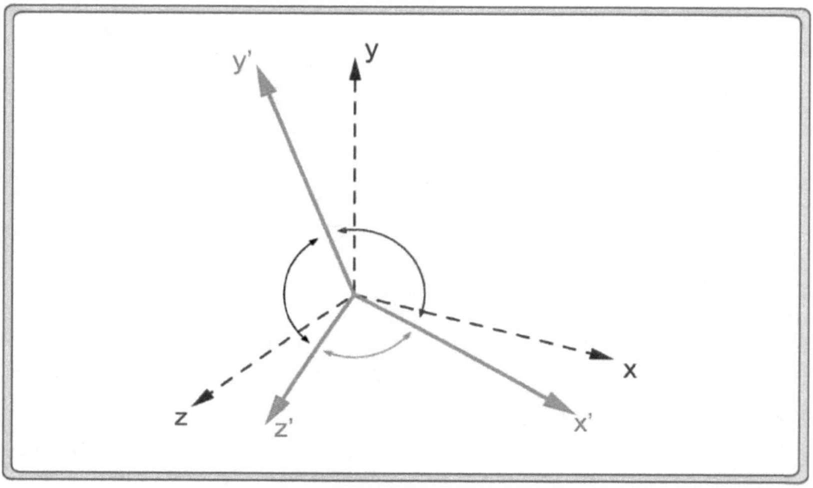

Figure 2.6: It can take time to solve everything with only a coordinate system with perpendicular intersecting axes. The oblique intersecting coordinate system should also be given a chance.

The shapes obtained in a coordinate system can have very valuable physical meaning. But mathematicians usually don't know this. In fact, it would be very good if they did, but unfortunately, they are not very curious and questioned about this subject.

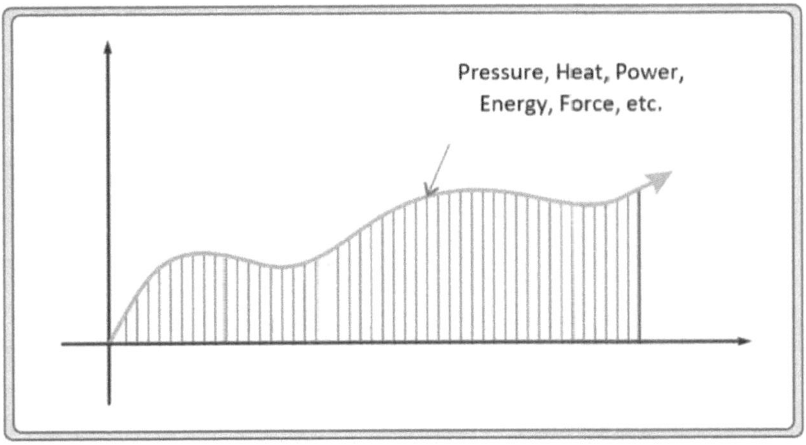

Figure 2.7: The same picture can mean different things.

You need to know that the area and volume of the shapes you transfer to the coordinate system, and of course the edges used to get to this area and volume, have a very valuable physical meaning. In fact, we spend our whole lives trying to figure out what they mean. This is actually the main reason why we solve geometry problems from primary school onwards. Unfortunately, because we get lost in the shapes, we only see the problems as geometry problems and don't investigate the truth behind them very much.

The area of a rectangle or a curvilinear region can physically and chemically correspond to many different expressions such as buoyancy, pressure, heat, energy, pH, number of molecules. Of course, a side that forms this area can also mean completely different things physically and chemically, just like the same area. To understand these, first pay attention to the units of formulas. Also keep in mind that these are concepts that are created by four operations!

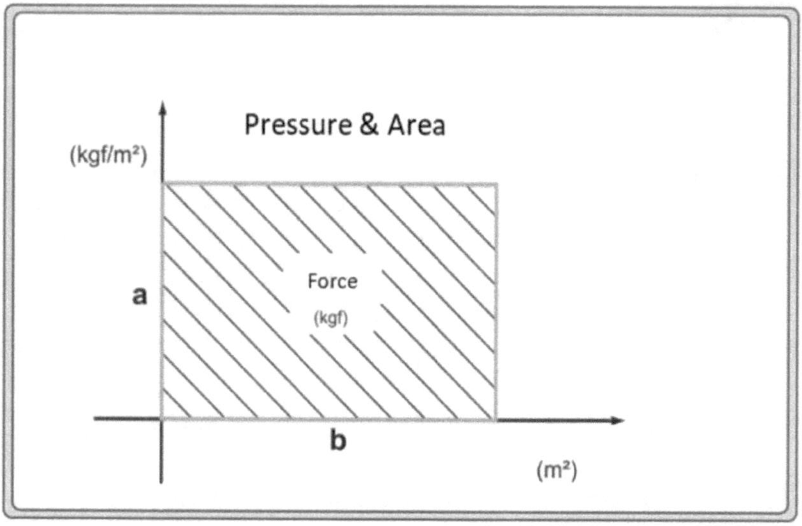

Figure 2.8: We struggle to find the meaning of the area and sides of a shape in a coordinate system.

You remember the formula for pressure, you get pressure by dividing the force by the base area, right? As you know, when you divide the area of a rectangle by the short side, you get the long side. Now if the area of the rectangle is the force, the short side of the rectangle can be the base area under the influence of the force. Then when you divide them together, the long side of the rectangle will of course be the pressure itself.

Mathematics, together with geometry, unifies all physical phenomena like this and makes them meaningful. You remember from the formula that you multiply the pressure by the base area to get the force. Mathematically, all you need to do is multiply the short and long sides to find the area of the rectangle. Of course, these operations are easy to understand with simple geometric shapes.

When the shapes you transfer to the coordinate system get a bit complicated, then the two most valuable functions of mathematics, the derivative and the integral, will help you. The coordinate system is the name of the place where all this is understood and where many of the mysteries of this life are revealed.

Real problems become mathematical expressions on the coordinate system. You start to live in a mathematical world. Sometimes this life lasts so long that you forget reality. You cannot leave the mathematical world and you start to perceive the geometric problem you solve as if it were a real problem.

Modern mathematics actually dictates this to us. It prevents you from seeing the truth. That's why I wanted to start by emphasizing that the coordinate system is the name of a scene in a theater play. Now let's start explaining the most important subject of mathematics, functions, without forgetting the coordinate system.

Functions

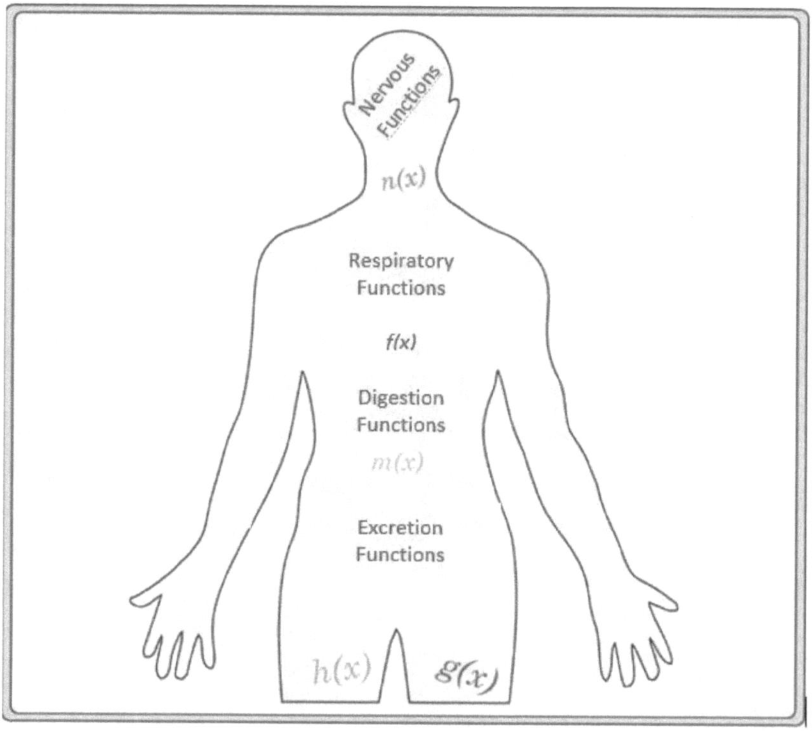

What is a Function and What Does it Do?

I would like to introduce functions by telling a memory. When I was in my first year of high school, we were studying functions. Three or four weeks had passed since the lessons. The school was a science high school and the students were curious. There were even discussions among ourselves, "Why are we learning this subject?" During one of the lessons, one of our friends asked a question, "Why are you telling us all this, what do we need functions for?" The year 1992, this was a question that perhaps millions of students at that time were eagerly waiting for the answer.

When we heard this question, of course, we were all focused on the answer from our math teacher. If we could get the answer to this question, we would start to form an idea about why the curriculum is the way it is at that age, and we would start to understand why other subjects were being learned. We would begin to question why we learn derivatives, why integrals, why limits, why trigonometry.

When we understood these, we would have learned the purpose of the education system. We would have realized at that age that memorization would not work and we would have regained our self-confidence that we had lost for centuries.

Trying to understand life through questions is perhaps the foundation of science. Think about how many questions you ask. Unfortunately, we are not very keen on asking questions and we keep postponing the questions that need to be asked to understand life. But we started asking the questions that everyone else left for last. If the answers to these questions had been answered properly back then, believe me, we would be a completely different country now.

We were all focused on the answer from the teacher, but the teacher did not give us an answer, but a homework assignment. Our teacher gave our friend a homework assignment to research why he taught this subject, and I say taught, and to explain it in class next week. I would like to open a parenthesis here: Unfortunately, in our culture, those who ask questions or have an opinion on a subject are always given a job related to the question they ask or the thought they interpret. I don't know exactly the reason for this behavior to prevent us from questioning life. I guess we all have such a defense mechanism to hide our own ignorance. That's why such an act came from our teacher.

Our friend's father was a math teacher and he immediately raised the issue with his father. A week later he received a reply from his father. It took a while for the answer to arrive. Because his father probably discussed this issue with his friends at school and they decided to cover it up with an answer that no one could question, know or object to. In the end, it was not the teachers' fault, it was up to them to answer the question "Yes, why do we teach these?" out of the blue.

Anyway, our friend stood in front of the board and explained that functions are taught to build rockets and space shuttles, they are used to build airplanes and satellites, they are used to travel to Mars and back, the Americans used functions to go to the Moon, in short, functions are taught to be used in space exploration.

While he was explaining these things fervently, our math teacher kept nodding his head up and down as if he had known this before and was confirming what he was saying. Happy to have given a research assignment, to clarify an obscure topic, and of course to explain such an important topic used in the construction of rockets and space shuttles, he continued the lecture from where he had memorized it. He was very happy and made our friend applaud for minutes, maybe even gave him an oral grade, I don't remember exactly.

As our friend explained, we were learning functions to explore space and build a space shuttle and we were very happy. We now know the answers to such questions. Later on, we were going to learn integrals and derivatives to explore space and build space shuttles. That's why we learned exponents and radicals. It was actually a good tactic! We were so happy... When faced with such questions, when you bring up space, no one dares to question whether you know the subject or not.

But what was missing here was the answer to the question "But how is it used?". Also, if everyone was learning this subject to build space shuttles, what was the need to train so many people to build space shuttles, and if so, where were they and our space shuttles? We did not have the answer to these questions. In fact, the answer was a nice answer to save the day, made up to keep people from questioning more. I didn't see this at the time.

Our teacher was not to blame for this. Because, just like his predecessors, he had graduated from university without knowing why he had learned these subjects and where they were used. Who did our education system expect this from anyway? Now, more than 30 years have passed since this incident, unfortunately, even now I see students in mathematics teaching departments graduating from schools and starting their professional lives without knowing why functions are learned and where they are used.

Students studying in education faculties take over the "task of graduating without knowing why they are teaching" from their teachers who graduated from those departments before them. Unfortunately, no one has the courage to say that the king is naked. I think the reason for this is that, as in everything else, we take the education system from the West without question. As such, only and only a rote education system emerges with unquestioned information. No one even feels the need to ask the question, "Why are these things being learned and taught?" If you make those who ask questions regret asking them, no one will criticize your system. That's why your math ranking remains more or less the same.

Anyway, about the effect of this memory on me... I was still happy with my friend's answer and I believed that there was a purpose for thinking so much, studying so much, and staying awake for days for the university exam. In fact, the study had paid off. In the university exam, I answered 50 out of 52 math questions correctly and I was entitled to enroll in the Department of Aeronautical Engineering at Istanbul Technical University, one of the departments where I would use functions in the best way. As you can see, I was very close to building airplanes and space shuttles.

I have always been complimented in math classes. But I want to say this clearly: I can say that I graduated from university without knowing how to use the mathematical expressions I learned throughout my education in any real problem in this life. And after graduation, I waited for years to use them.

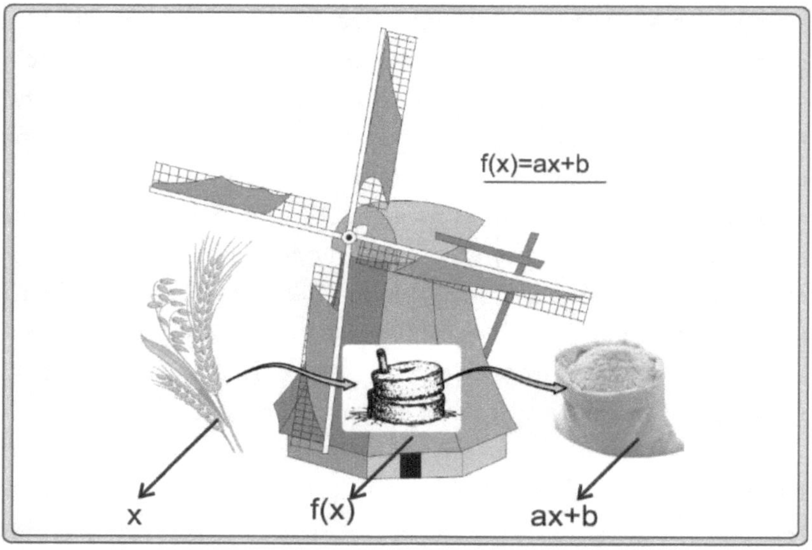

Figure 3.1: The mystery of life is hidden in functions.

Of course, functions are often used in space exploration and space shuttle construction. However, defining functions as "they are used in space exploration and space shuttle construction" is not only an incompleteness. It also opens the door to the great mistake of obscuring the knowledge of a subject. Unfortunately, such answers are given by people who do not fully understand the functions in order to avoid being banned.

Figure 3.2: Everywhere we see functions.

Functions are one of the most basic mathematical functions that help you transfer any event we encounter in life to paper. From the transformation of wheat into flour and meat into minced meat, to the transformation of the food we eat into urine and feces, to the change and transformation of the air we breathe in our bodies, to the events of photosynthesis and chemosynthesis, to the growth of a plant by feeding it with light, water and soil, to the fruiting of a plant, whatever situation, event or relationship there is in our lives, we can transfer it to paper and analyze it all thanks to functions. If you observe properly, you can find and extract functions in everything. You don't learn the functional relationships in the frog's digestive system so that you can intervene when something happens to it. You learn the functional relationships there to add depth to your analytical perspective. You may ask what function has to do with the transformation of meat into minced meat and how we can see the functional relationship in this case. You need to look at it as a volumetric change. Meat has a shape that changes before and after it enters the meat grinder, right? When the meat comes out of the machine, it turns into thin pieces like pasta sticks. The shape of the wheels inside the meat grinder gives you the final shape of the minced meat. If you play with the shape of the cogs, you can change the shape of the minced meat. The main thing here is to connect the volumetric change of the meat with the shape of the wheels. If you put the edges and corners you use to define volume into numbers and create a table between before and after, you can reach the functions from there. You can look at the transformation of wheat into flour in the same way. Remember, you are not changing the weight, you are changing the shape of the substance, and the functional relationships are in it! Don't be fooled by my simplicity, you need to do a lot of coordinate transformations and mathematical operations to reveal these functional relationships.

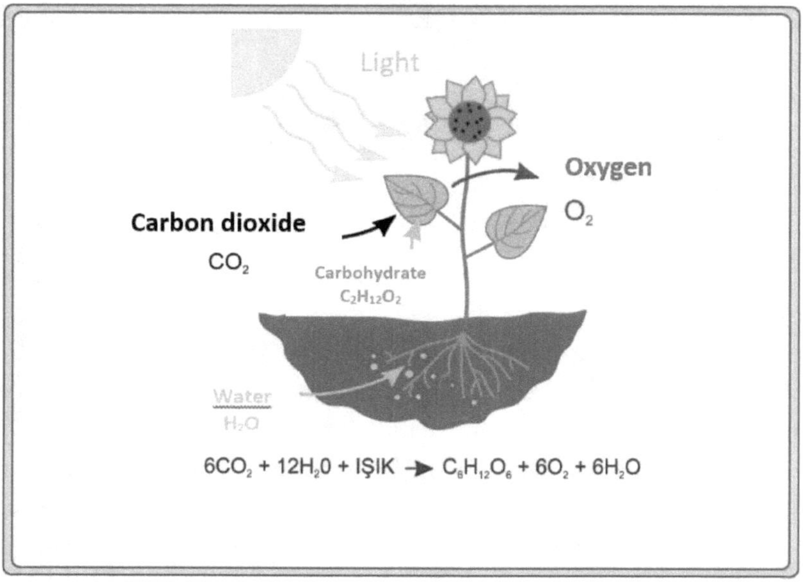

Figure 3.3: All facts of life make sense with functions.

From the turning of the steering wheel of a car to the movement of the tire, from the pilot's touching of the elevator to the airplane's turning left and right, every event is of interest to functions. If there is change and transformation anywhere, we can talk about a functional relationship. A painting or a sculpture also emerges as a result of a functional relationship. From this point of view, you can also call the mathematical function that defines the relationship between the real image and the virtual image a function.

In the equation $f(x) = ax+b$, which is the most basic functional representation, you can consider that one side represents the real world and the other side represents the virtual world. In this respect, you see that functions are a concept produced as the most basic mathematical function that acts as an operator to transfer the events in the real world to the virtual environment you fictionalize on paper. Of course, not only change and transformation are explained by functions, but every formula is actually the name of a functional relationship. Expressions such as $F = ma$, $E = mc^2$ are of course functions.

Functions should not only be seen as concepts that establish the relationship between the real and virtual world. You may also need to use functions in transitions between virtual worlds. In the future, functions will again be the key word for people who want to travel between parallel universes and live there. In short, whatever situation, event or relationship there is in this life becomes visible and can be put on paper thanks to functions.

Take functions out of mathematics, you will need another expression to understand the facts of life, and whatever you find will probably look like functions again. When you take functions out of your life, you cannot put real events on paper. How do we solve problems that cannot be put on paper? That's why you need a transducer like a function to put problems on paper and to make expressions meaningful.

Everyone with a high school education has learned that $f(x) = x^2 + 3 + 8$, He encountered different functions such as $f(x) = 3x^4 + 6x^2 + 4x + 7$, $f(x, y) = 4xy^2 + 8x^3 + 6y + 5$, and drew graphs in the coordinate system using them. The biggest problem in transferring real life problems to paper is defining x and y and finding the coefficients in front of them. We also have the problem of not immediately grasping why they are multiplied, added, divided or subtracted.

Think of an airplane engine, you think of a very complicated system, right? But if you know math, you will see that it is actually not that complicated. There may be thousands of functions, but they are all derived from the most basic function, which is to increase the speed of the incoming air and push it out. Math tells you that. But unless you do real experimental work to build an airplane engine yourself, of course you can't see the math behind it.

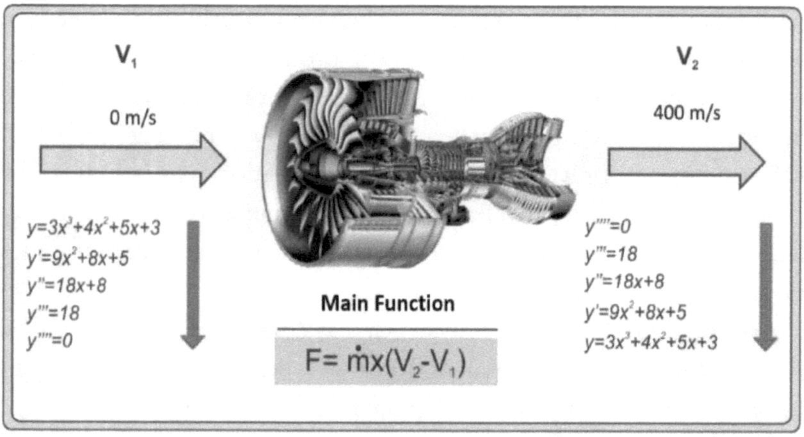

Figure 3.4: There are thousands of functions inside the engine in an airplane.

The root of this problem is that we cannot create our own problems. So, when we solve them on paper, we don't own the results as much. We necessarily need to look at the answers. Unfortunately, in most experimental studies, we draw graphs adjusting the results according to what should happen.

Now let's move from the airplane engine to simpler functional relationships. For example, how can we relate the number of rotations of an automobile wheel to the distance it travels? How can we define a relationship between the number of rotations of any gear that turns that wheel and the number of rotations of the wheel? Likewise, how can we reveal a relationship between the angle of rotation of the steering wheel and the angle of rotation of the wheel? What is the relationship between the rotation of the piston in the cylinder and the rotation of the wheels? You can multiply questions like this on the way to function.

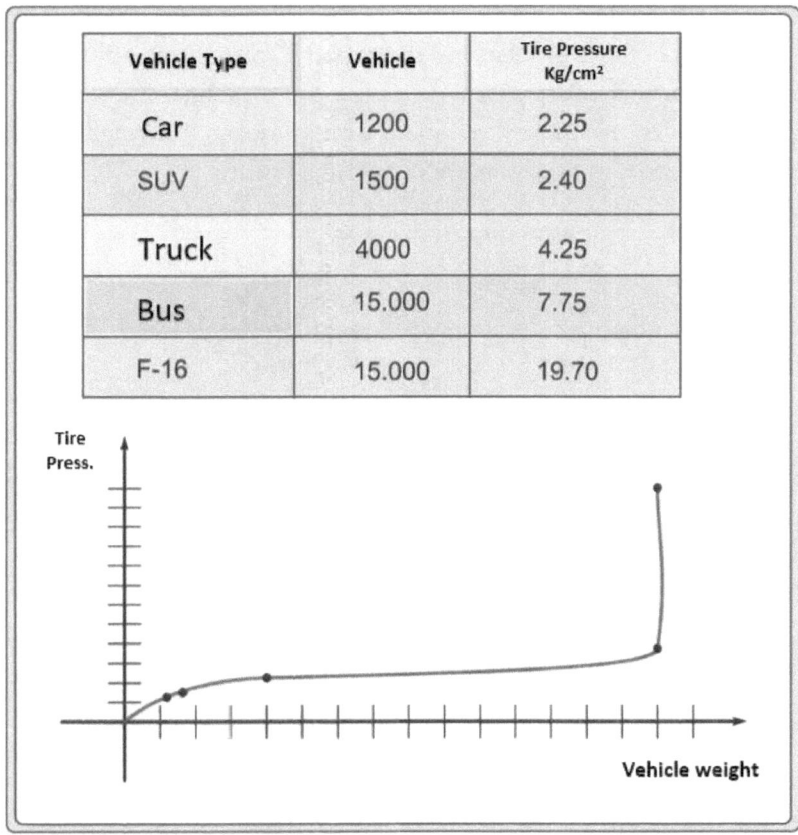

Vehicle Type	Vehicle	Tire Pressure Kg/cm²
Car	1200	2.25
SUV	1500	2.40
Truck	4000	4.25
Bus	15.000	7.75
F-16	15.000	19.70

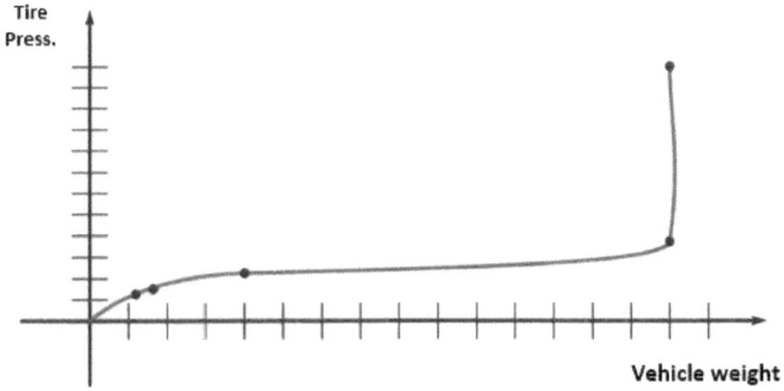

Figure 3.5: Life does not always present linear relationships.

Why not also define a relationship between the weight of a car and the tire pressure? Such problems are not difficult, of course, and most of them, in their simplest form, aim at establishing a functional relationship such as f(x) = 3x, f(x) = 4x, etc. For example, in the problem about the number of rotations of a car tire and the diameter of the tire, if you know the circumference of the tire, say 2 m, then of course the tire has to rotate 500 times for a distance of 1000 m and you can define the function as f(x) = 500x.

Considering the tire pressure of a car weighing 1500 kg and the tire pressure of a car weighing 2500 kg, what relationship can you define between car weights and tire pressures? If you define these relationships properly, you can even find a general formula that says, "What should be the tire pressure if the car weighs what?". You will be very surprised to see that you can't have proper linear relationships here. If you define this relationship properly, you will actually see that many functional relationships in this life are defined curvilinearly. Even the relationship between the tire pressure and the weight of the car will tell a lot to eyes that see everything linearly.

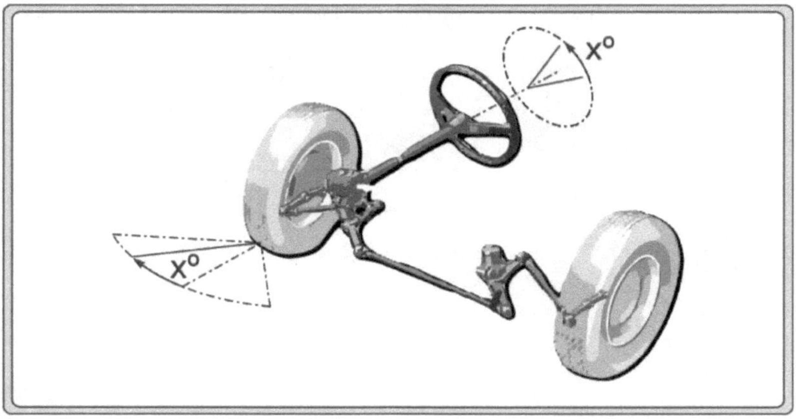

Figure 3.6: You can see functional relationships everywhere you look.

You can also define an angular relationship between the rotation of the steering wheel and the rotation of the wheel. Here the rotation angle of the wheel can be f(x), x can be the rotation angle of the steering wheel, and of course many linear and non-linear relationships can be established between them. If it is a linear relationship, you can get away with a definition like $f(x) = 100x$, that is, with only one coefficient. Now don't get confused when I make you define many such physical relationships. In fact, in order for you to comprehend and understand these functions in depth, I want you to internalize this subject with the events you see and constantly witness.

There are plenty of non-linear relationships in life. Aren't most of the functional relationships we encounter in life already built in multiples of ups, downs, stagnation? So, don't always expect linear variability from this life. If there is a non-linearity, i.e. the steering wheel turns 10° and the wheel doesn't turn at all, then the steering wheel turns 30° and the wheel turns 12°, then the steering wheel turns 60° and the wheel turns 20°, you can create a data set. From this you can draw a graph and from this graph you can get a function. In short, we are trying to explain to you here that it is not very difficult to establish a functional relationship between the steering wheel and the wheel in a very simple way rum.

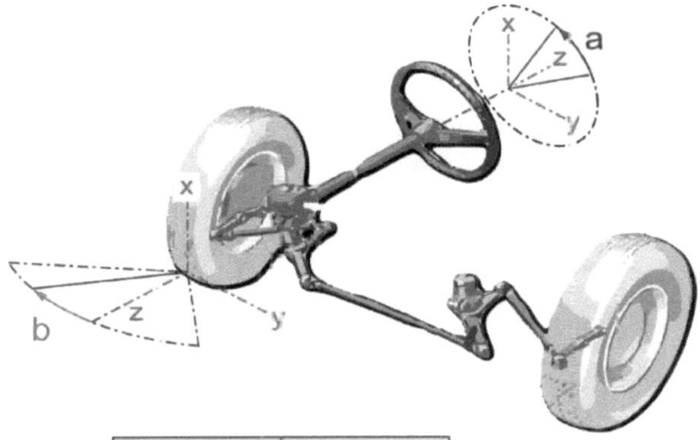

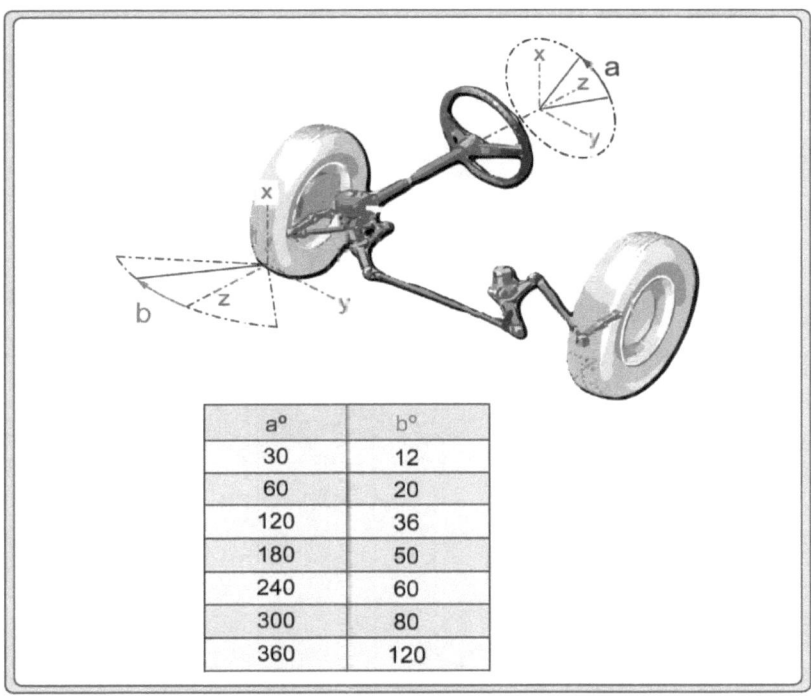

a°	b°
30	12
60	20
120	36
180	50
240	60
300	80
360	120

Figure 3.7: Mathematical concepts come to the rescue to draw meaningful conclusions from non-linear relationships.

When I put the values I just mentioned into numerical expressions; I got a function like $f(x)=0.0008735x^2 -0.0571x+24.46$. Of course, here you may need to limit the independent variable x so that it does not go to infinity. For example, you could define a range so that x takes a value between *0* and *360*. From here you should also be able to see this: You can go from the measurements to the graph and from there to the function. If you want to define the angle relationship between the steering wheel and the wheel using the data, you can also get the graph and from this graph you can get the function. It is useful to emphasize this one more time: Even simple examples like this tell us that everything we learn as a formula is actually a special function and that almost all of them are formulated thanks to graphs created as a result of experimental data.

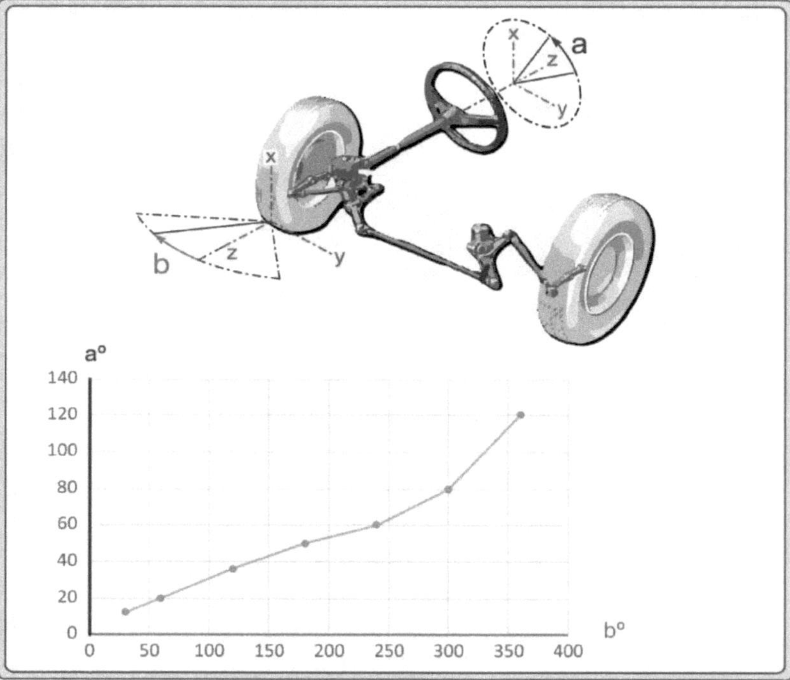

Figure 3.8: Measure, create a table and draw a graph, and the resulting figure will help you to see the reality.

There are so many functional relationships in our lives, maybe millions, and this number is constantly increasing. Because you can define functional relationships for every system, every situation where there are inputs and outputs. So it is very difficult to determine this number. Millions of relationships can be defined not only in science but also in social sciences.

We can talk about functionality in many areas, from the temperature of the weather and the change in our emotions, to the speed at which children learn to speak and the number of people in their upbringing, to the relationship between the change in the currency and the level of education. So, don't worry, "Europeans and Americans have found and defined everything, there is nothing left for us!" Anyone can find many functional relationships in this life. Such relationships should become so commonplace for you that then you will understand this life deeply and start solving problems.

Even at home you can analyze so many simple events that it will open the door for you to use functions to grasp many issues in this life. For example, weigh yourself with your partner, you and your child and record your own weights, then take an object of a certain weight and have everyone hold it parallel to the ground and record the time it takes for each person to lift it. Try to establish a functional relationship between your own weight and the time it takes to lift the weight. Don't give me excuses like "everyone has different musculature", just do as you are told and try to get graphs. Then you will have a better understanding of what functions mean in this life and you won't bring the subject to the space shuttle.

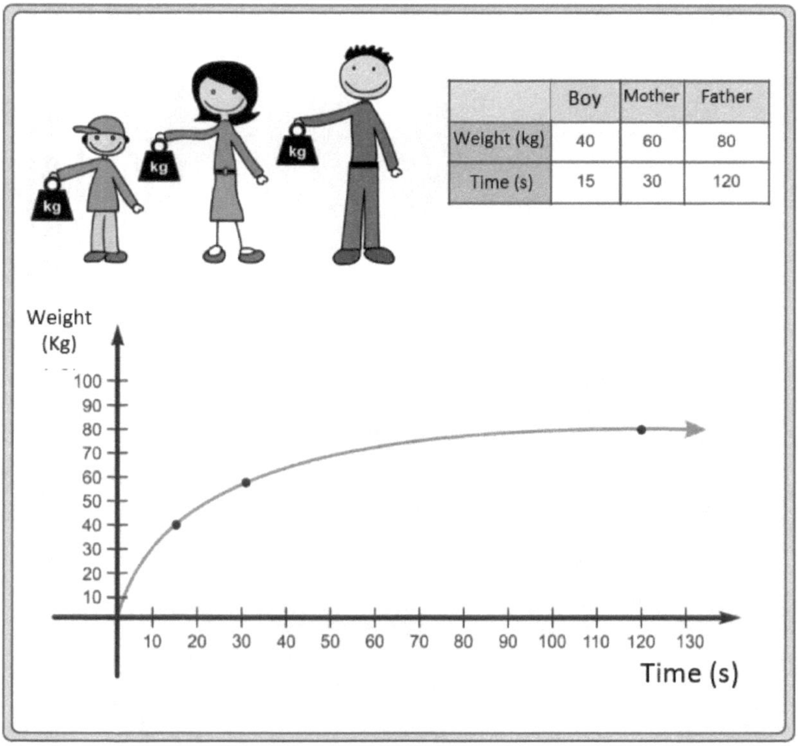

	Boy	Mother	Father
Weight (kg)	40	60	80
Time (s)	15	30	120

Figure 3.9: Even the inside of your home can tell you a lot about functional relationships.

More or less functional relationship is about how you live this life. If you are living this life to fulfill your basic needs like eating, drinking, shelter, nobody will ask you, "How many functions do you have?" If you want to make a difference and solve all the mysteries of this life, then you really need a lot of functions. You can solve the problems in our lives with the equations you set up as a result of functional relationships.

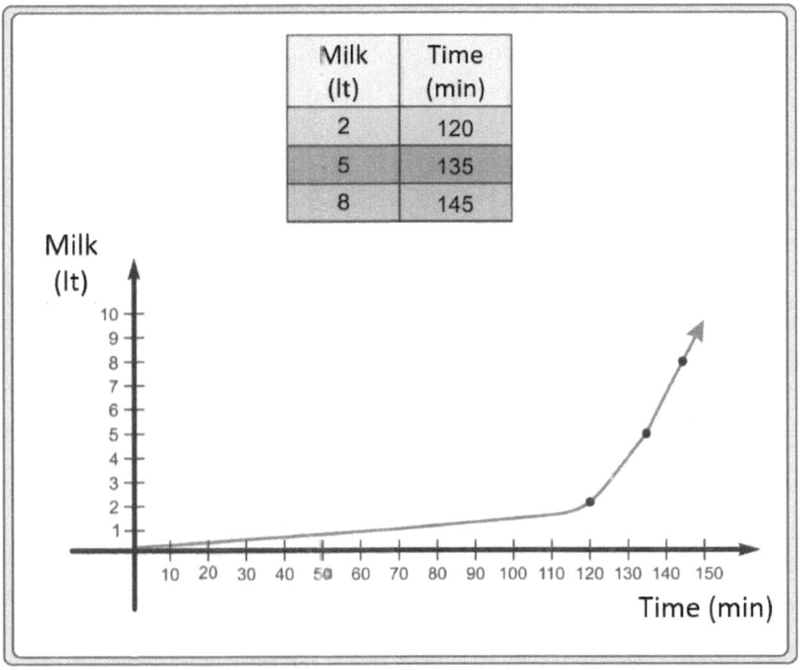

Figure 3.10: If you don't choose the right units, your image will become cluttered.

I recently bought 15 liters of milk to make cheese and yogurt. After boiling the milk, I poured it into 2-, 5-, and 8-liter containers. Then I waited for the milk to cool down to 45°, the ideal fermentation temperature. I waited for the milk in each container to cool from 100° to 45° at different times. How long do you think I waited for each container?

I waited 2 hours for the milk in the 2-liter container, 2 hours and 15 minutes for the milk in the 5-liter container, and 2 hours and 25 minutes for the milk in the 8-liter container. Now, for eyes that see everything linearly, if the milk in the 2-liter container reaches 45° in 2 hours, the milk in the 5-liter container should reach that temperature in 5 hours, and the milk in the 8-liter container should reach that temperature in 8 hours. So where did these times come from, right? This life often does not allow you to establish simple linear relationships. Here functions tell you that the problems you face in life often involve curvilinear relationships. I can say that the mystery of this life is hidden in curves. In fact, change the units, redraw the graph using seconds instead of hours and minutes, and the shape of the curve immediately changes, doesn't it? For those who say, "What are the numbers e and π for now?", I answer: We make these curves meaningful with the numbers e and π.

Defining functional relationships depends on your level of knowledge and how you want to approach the problem. If you know little about it, you define few functions; if you know a lot, you define many functions. For example, you can define dozens of problems in the construction of a car, from acceleration to weight and volume, from acceleration to braking distance, and write functions to make sense of them. But if your level of knowledge is much higher, you may need to define functions on many different topics, from the material structure of the car to the problems of frictional heating. If you want to go even deeper and see the effect of engine vibration on the passenger or passenger doors, you can define functions that will reveal these as well. The more detailed and deeper you approach the problem, the richer your functions will be. Take a look at the development process of automobiles; while the first automobiles were designed only to carry people, that is, they had no other features other than that, think of the features of the new generation of automobiles as people's demands increased. You can clearly see how the number of functions has increased and can increase, even on just one automobile.

I want to show how the functional relationship deepens with a simple example. For example, you have an iron rod and you want to determine the maximum load it will carry by putting a weight on it and observing its deformation in the horizontal plane. For this, let there be a limit to the amount of inclination accepted in the iron bar. Now you can reach this figure by simply increasing the weight. Here you can get a figure at a constant temperature, i.e. independent of temperature. You can reach the maximum load so that this bar can withstand only this much load. If you see that a different value is reached under temperature, then you need to add a temperature variable to the equation in addition to the load variable. Here is an example showing the transition from an equation with one unknown to an equation with two unknowns. As you can see, as the number of variables increases, the depth of your problems increases and you will reach more realistic results.

You can write the amount of bending in the bar at constant temperature under standard atmospheric conditions as a function of the elastic modulus. Writing it as a function of the elastic modulus means that the variable is the elastic modulus. If you write the change of the elastic modulus as another temperature dependent function, then the subject will be even deeper.

Normally bending $\delta = \frac{PL}{AE}$ While it is calculated with the formula, by considering the temperature as a variable of the elastic modulus, the flexural $\delta = \frac{PL}{AE(T)}$ is calculated with the formula. Another variable inside another variable, here is a good example of nested functions! As your perspective deepens, the number of nested functions will increase. For example, you can write time as a function of altitude...

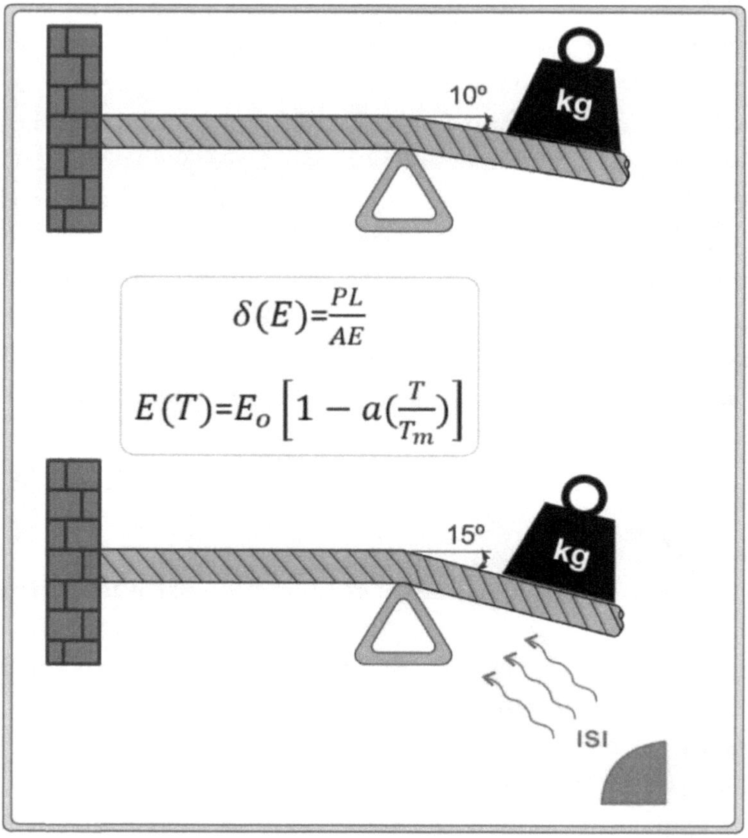

Figure 3.11: There is a function inside a function.

$$(E(T))= E_0\left[1 - A(\frac{T}{T_m})\right], \delta = \frac{PL}{AE(T)}$$

P= Load,

L = Length of the rod,

A= Cross-sectional Area of the Bar

E_0 = Elastic Modulus at 0 Kelvin Temperature

E(T)= Elastic Modulus at Temperature T

T_m =Melting Temperature of Material

δ=Bending Amount

Variables can sometimes enter into the functional relationship we call an equation here, sometimes by multiplication, sometimes by addition, sometimes by division or subtraction. This is related to how you handle your data set and often occurs when analyzing experimental data. Here the units can enter into your functional relationship even as coefficients. For example, the gravitational constant in the law of gravitation differs depending on the system of units you use. Just changing your coefficient will change the shape of your graph. So, no matter what, stick with whatever system of units you are using. Do not try to solve problems using a different system of units, you will get confused. You cannot touch your own life with a different unit system.

Quantity	SI Base Unit	English Unit	Derivative Unit
Length	Meter (m)	Feet (ft)	1ft:0,3048 m
Mass	Kilogram (kg)	Libre (Pound) (lb)	1lb=0,454 kg
Force	Newton (N=Kg*m/s^2)	Pound-force (lbf)	1lbf=4,45 N
Energy	Joule (J=N*m)	Food-pound-force (ft*lbf)	ft*lbf=1,356
Power	Watt (W=J/sn)	Food-pound-force per second (ft*lbf/s)	1ft*lbf/s=1,356 W
Pressure	Pascal (N/m^2)	Pound per square inch (psi)	1psi=6.895 Pascal

Figure 3.12: Do not expect to get the same expressions with different unit systems, even if they mean the same thing.

I tried very hard not to make any mistakes when creating the table above, because multiplication and division can get confusing. Add to this the difference in coordinate system graphs and you can see how difficult it is to solve problems with different unit systems. So, try to read things in your own unit system. Now look at the unit of the gravitational constant and you will see that it corresponds to a completely different set of numbers between the international and the British unit system.

In the literature, the most precise value of the gravitational constant G measured with the international system of units (SI) is given as $6.674*10^{-11}$ m /kgs^2. Now, when you convert meters to feet and kilograms to pounds and write the equivalent of this constant in the British system of units, you reach a value like $0.42*10^{-11}$ ft /Ibs^2. As you can see, the gravitational constant corresponds to completely different groups of numbers even in two different unit systems.

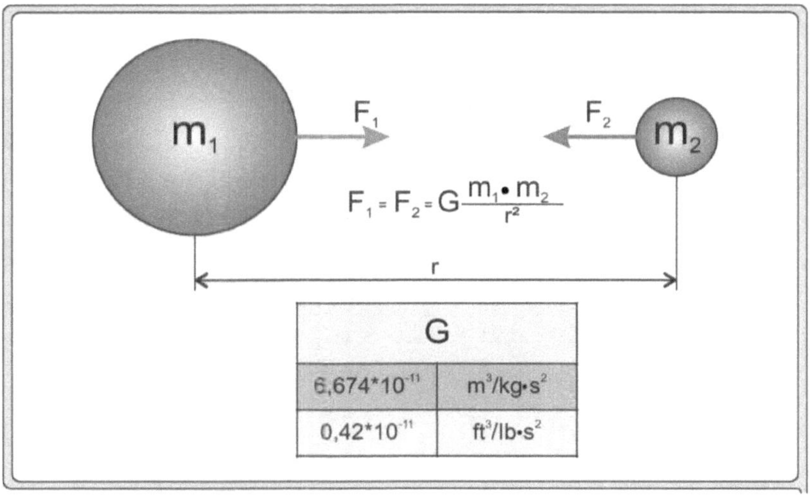

Figure 3.13. You get different expressions with different units.

Similarly, we use 9.81 m/s² for the acceleration of gravity. In the British unit system, this value is 32,174 ft/s² as you can see, we convert them to each other using a function like y=ax. In summary, we use functions even when switching to different unit systems.

Mathematics is the art of defining problems with letters, numbers and symbols. In other words, mathematics comes into play as far as you can define it. If you can define functional relationships down to the smallest detail of a problem, the solution to that problem will of course be the closest to reality.

Looking at the formulas defined so far, it is actually possible to define all relationships in life as first, second or at most third order functions. Adding and multiplying them yields functions of much higher degree. So, when you see equations of the fourth degree and above, don't say "Where did that come from!".

In conclusion, I have tried to explain that functions have more meaning than building space shuttles and using them in space exploration, and that they have a meaning in this life. Even if you just look at the formulas, you can see the picture more clearly. The world of formulas actually explains very well why you learn functions, doesn't it? I tried to emphasize that you need to define the problems you face in life and that functions are the most basic mathematical functions for this definition. Therefore, you can see and use functions as the most basic mathematical expressions you will encounter at the entrance gate of the analytical thinking system.

Trigonometry

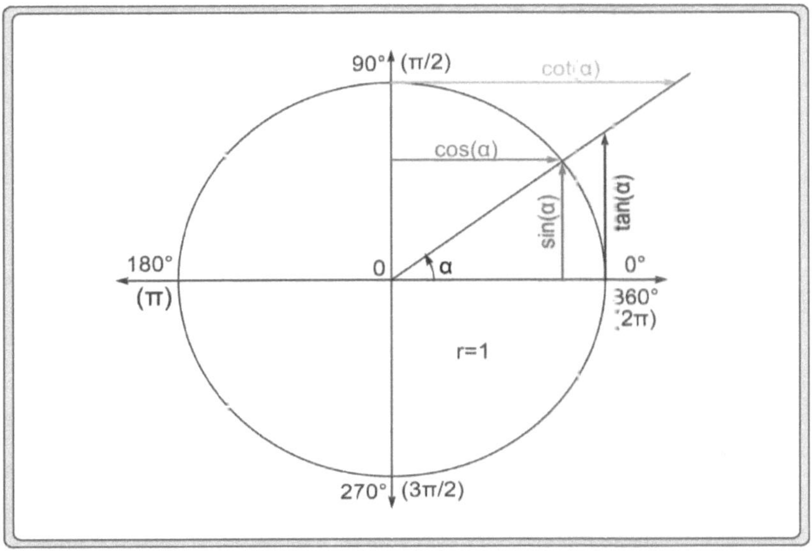

How Trigonometry Originated
and
Why are we learning about it?

The history of trigonometry dates back to prehistoric times. I don't know what comes to mind when you think of trigonometry, but I immediately think of triangles. Actually, this is normal, because "tri" means 3, "gon" means edge, and "metry" means metric measurement; in other words, trigonometry is a word that means three-sided measurement. If its name was in Turkish, perhaps this subject could be understood much more easily and learned immediately. Since it has entered our lives as it reads from Latin, when one hears the name trigonometry, one hesitates and feels like saying "This subject cannot be simple, it has to be difficult!".

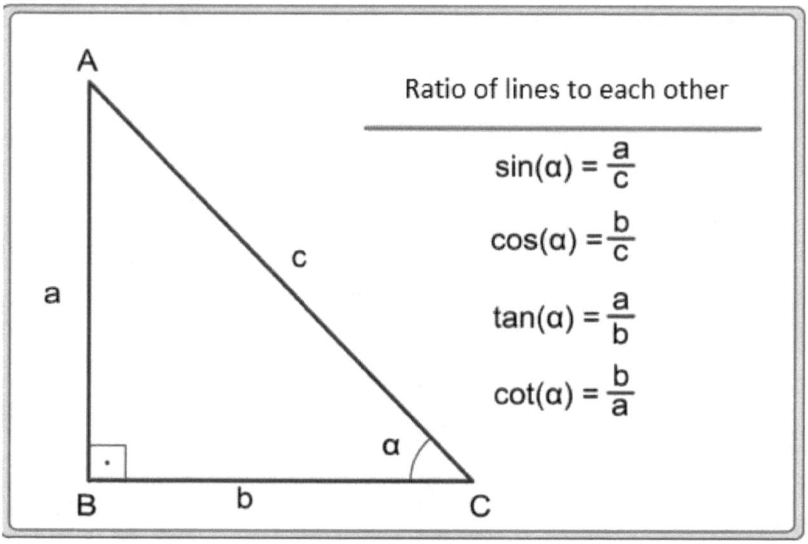

Figure 4.1: Sine and cosine expressions are line segments resulting from proportion.

Trigonometry stands out as a mathematical function enriched with functions such as sine, cosine, tangent, cotangent, in addition to the calculations we make on the triangle. Trigonometry has been part of the curriculum from secondary school onwards, and it manifests itself very seriously in all aspects of further academic education.

Scientists have achieved very good solutions and results when they combine geometry and mathematics. Looking back at my education life, I solved a lot of math and geometry problems. Especially solving angle and side problems in triangles was my biggest hobby. When I solved them, I would be very happy and say, "I am glad that math and geometry exist and I am glad that I am learning them!"

However, years later, I realized that if you don't know where these concepts come from and where they are used, the problems you solve are no different from solving puzzles. Nowadays, I say to myself, "I was solving puzzles back then." Because I solved so many questions, but I never needed them anywhere in my life and therefore never used them. The number of people who use them in our country is so small that even though I am so interested in education, I can't help but ask, "What would happen if they were not learned?" We are a country that produces and solves technician-level problems rather than engineering-level problems. That's why we don't even feel the need to actually use trigonometric functions, which are almost everywhere in a real engineering problem. There has been some movement in this regard recently, but not enough.

Puzzles are mostly known as activities that retired people solve to pass the time. Don't get me wrong, puzzles are important and develop the mind, so keep solving them. However, if you want to build an automobile, an airplane, an engine, a bridge, a dam, a combine harvester, a skyscraper, a subway, a telephone, a computer, a satellite, you need to know why the most basic concepts of mathematics are taught and learned. If you think you will get anywhere by memorizing rules, you are wrong. That path will only lead you to the old people's club solving puzzles.

This life contains many concrete examples of how circular motion can make things easier. From the invention of the wheel to the invention of the engine, we have experienced and are experiencing this. Take a look at human history. Look at the rate of technological development from the invention of the wheel to the invention of the motor, and then the rate of technological development from the motor to today. Compare these three periods and you will better understand what I mean.

Figure 4.2: The power of circular motion.

As you can understand from the rotation of the earth, circular motion holds the most important secret of this life. The nirvana of geometric shapes such as triangle, rectangle, pentagon, hexagon is the circle. If you analyze the path from triangle to circle well, it will be very easy for you to solve many problems in this life. Here I am actually telling you that trigonometry was discovered to reveal circular movements by using triangles, that is, lines.

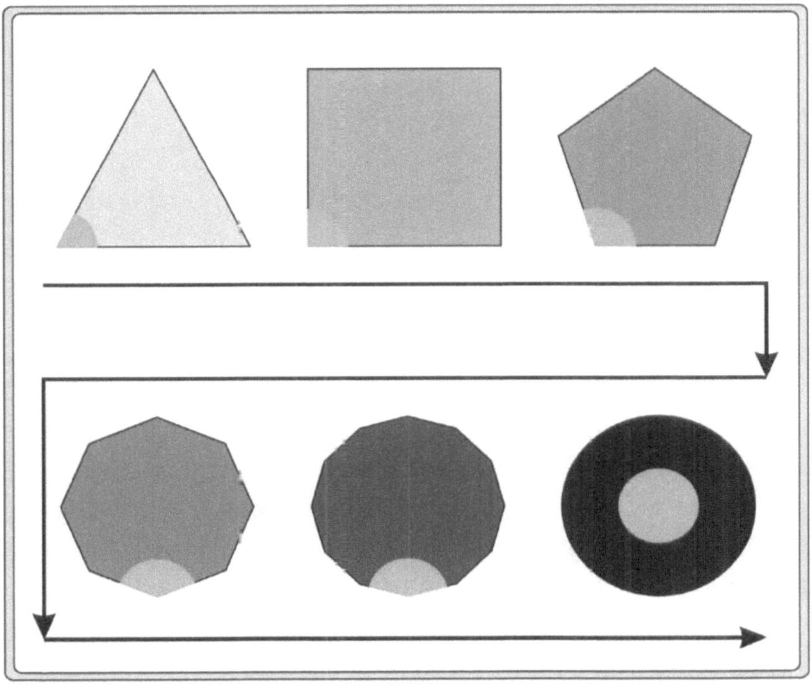

Figure 4.3: From triangle to circle, trigonometry always helps you and helps you see the truth.

To understand trigonometry, you first need to know what concepts such as sine, cosine, tangent and cotangent mean. Mathematicians have defined sine as the ratio of the opposite side to the hypotenuse, cosine as the ratio of the adjacent side to the hypotenuse, tangent as the ratio of the opposite side to the adjacent side, and cotangent as the ratio of the adjacent side to the opposite side in a right triangle.

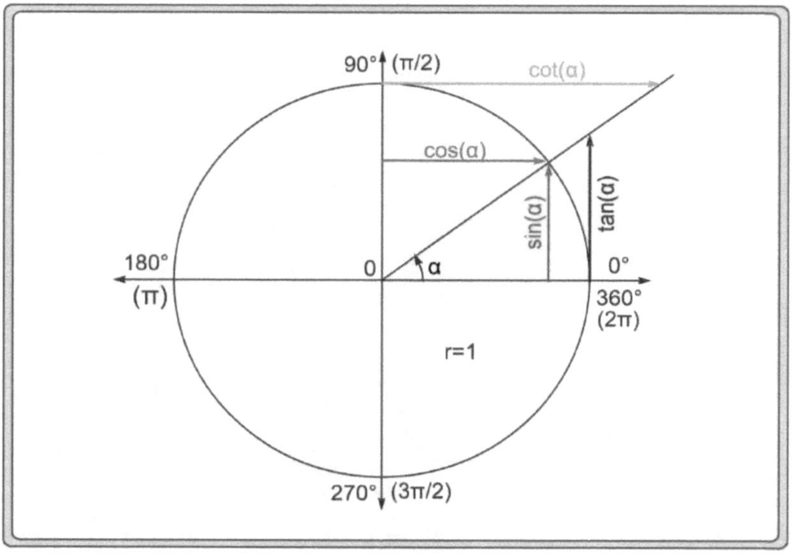

Figure 4.4: The unit circle shows reality in all its nakedness.

These concepts in trigonometry are explained in terms of the unit circle in which a triangle with a radius of 1 unit, i.e. a hypotenuse length of 1 unit, is placed. Since the result does not change when the opposite or neighboring side is divided by 1, there is no need to calculate the hypotenuse in unit circle equations. This notation is intended to make it easier for you to see and understand the picture with the right sides of the triangle. As you know, thanks to functions, we can put the facts of this life on paper.

It is not important to make functional equations too long. The important thing is to be able to write equations where you can get the result quickly and easily. This is why equations in the unit circle are explained. The hypotenuse being 1 gives you this convenience. Concepts in mathematics are created and derived so that you can do this kind of analysis. For example, in equilibrium problems $\Sigma F = 0$ and ΣThe equation $M = 0$ is used. In airplane design, problems are solved based on the fact that the net force resulting from the balancing of the carrying force and the weight of the airplane during level flight is zero. Look at the simplicity and simplicity of the equation, and remember that this simple expression paved the way for the airplane!

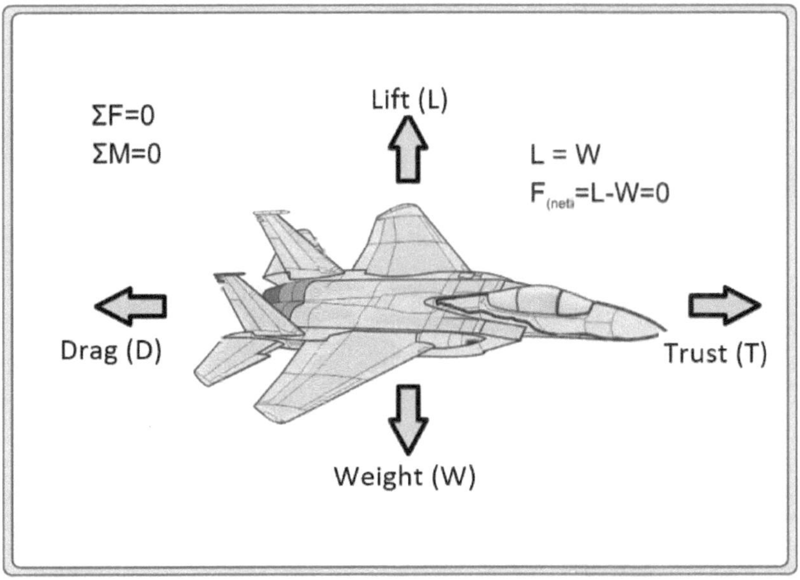

Figure 4.5: With equilibrium equations you start from scratch.

Of course, if you want to understand topics like trigonometry better, you need to know the meaning of basic terms, otherwise you cannot communicate with the problems. Just as everyone has a name, the concepts produced and to be produced in mathematics have to have a name. However, I would like to emphasize in a separate parenthesis that there is a need for a unique point of view on symbols and notations, and therefore more memorable symbols and notations that appeal more to our point of view should be derived.

Don't get too confused by the fact that in mathematics terms are constantly being coined and we keep converting them into each other. Focus on the picture rather than the terms. Pictures tell you more than symbols, letters and numbers.

We have already defined sine and cosine and you need to master their functional representations. You will understand better in the following chapters, sin(x) and cos(x) functions represent very valuable transformation functions for you. Regardless of the value of "x", when you look at their outputs, you will see that they take values between -1 and +1.

Now I would like to show how mathematical transformations arise through a few small examples. We use these transformations so much in integration and differentiation that you can appreciate them even more. Let's look at how a few important trigonometric transformations arise, using two simple figures:

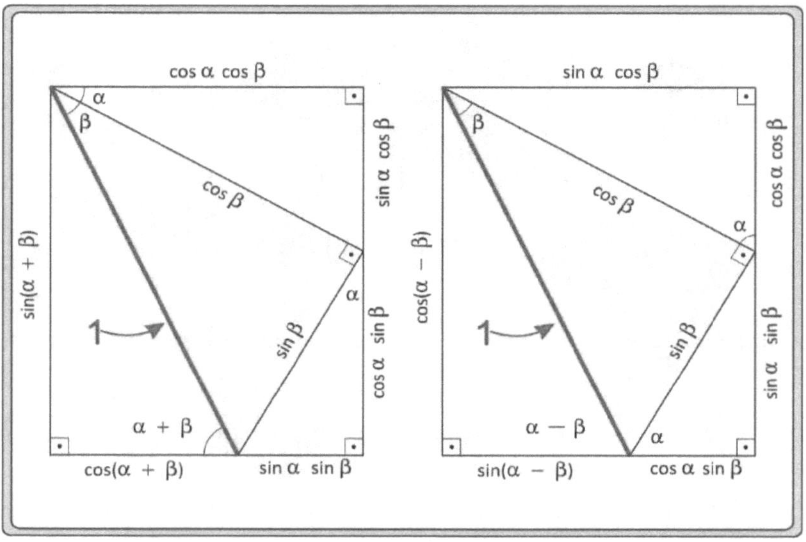

Figure 4.6: Sine and cosine are line segments, so doing four operations with them is inherently easy.

If you prove these equations, you can perceive that other trigonometric transformations follow a similar logic. You can deduce them later. First, I want to show the above equations on rectangles and triangles:

$$\sin(\alpha + \beta) = \sin\alpha \cos\beta + \cos\alpha \sin\beta;$$
$$\sin(\alpha - \beta) = \sin\alpha \cos\beta - \cos\alpha \sin\beta$$
$$\cos(\alpha + \beta) = \cos\alpha \cos\beta - \sin\alpha \sin\beta$$
$$\cos(\alpha - \beta) = \cos\alpha \cos\beta + \sin\alpha \sin\beta$$

You can see how these equations arise when you look at the picture on the previous page, right? If not, take a careful look at the opposite sides and you will see it immediately. Of course, the triangle with a hypotenuse length of 1 unit has a great contribution to this equality. When you move it to the unit circle, you get more equality. You can easily see that expressions with sine and cosine mean length when you divide the opposite or neighboring side by 1. So appreciate the unit circle. I used to say, "What is the need to explain sin(x) and cos(x) functions so much?" When I saw how much they are used in engineering problems, I realized how wrong I was.

Now, if you examine the shapes produced by the rotation of a point on a circle, you will see the path from here to trigonometry. In most of these figures, you will see that the triangle and the circle are used. You can encounter almost all of these shapes in static and moving systems. Especially in rotating systems such as motors, gears, screws and many mechanisms, you will see them frequently. After electromagnetic waves and the discovery that light moves in waves, it goes without saying how useful trigonometric functions are.

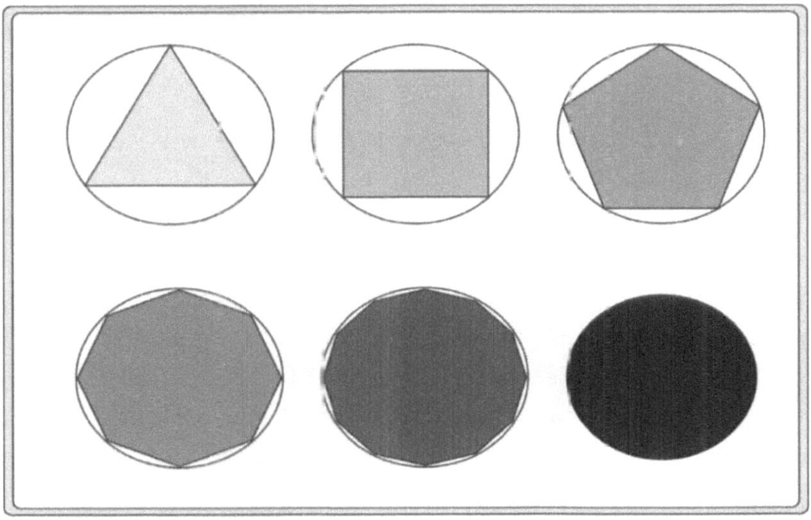

Figure 4.7: Trigonometry is another name for the relationship between a triangle and a circle.

Do you have a wardrobe at home and the door does not fit properly? Observe the movement of the door when you tighten or loosen the screws at the hinge points. Doesn't a door that snaps into place and closes properly with one turn of a bolt seem very interesting to you? Here you see the functional relationship between linear and circular motion even in a very ordinary event in daily life.

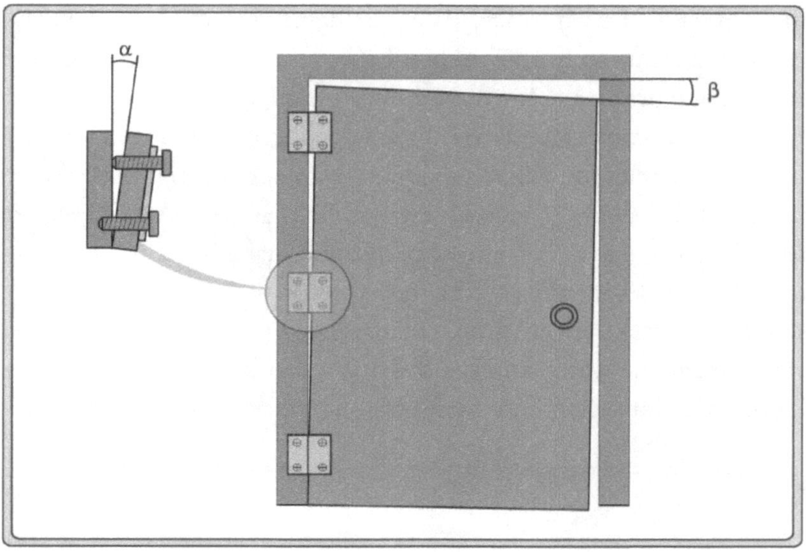

Figure 4.8: The change of angles creates the curve.

As you can see in this example, don't expect functional relationships to be linear anymore! Even children can solve linear functions. Move away from the elementary level and build complex functional relationships and deal with solving them. Just imagine that you will find a lot of functional relationships. Your finding and solving more complex non-linear functional relationships are the main goal of mathematics teaching. Otherwise you don't even need to learn so many mathematical concepts!

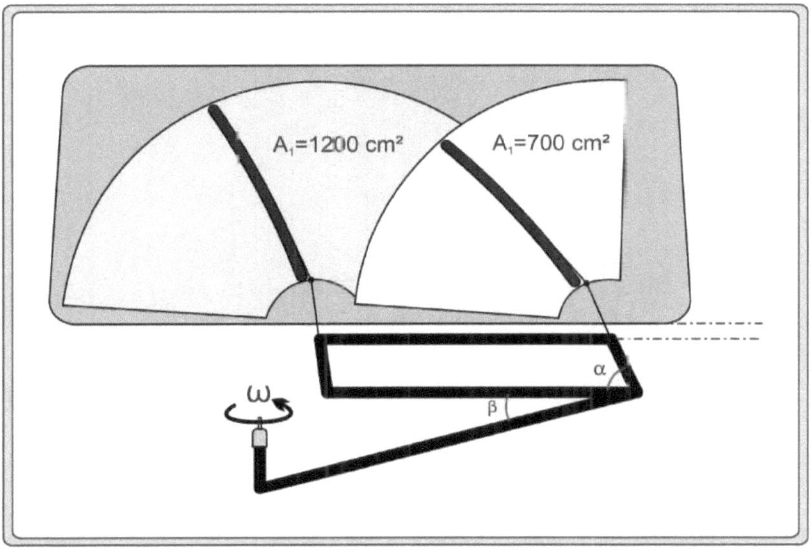

Figure 4.9: Where mathematics meets geometry, there is trigonometry.

Nowadays, we have all traveled in cars. Have you ever wondered how their windshield wipers work? An arm that rotates on a circle, rods that draw a parallelogram-shaped picture with the rotation of the arm, and wipers on top of them that wipe the glass... Almost all motor vehicles use this mechanism. Is it very hard to find? Absolutely not! We just need a careful eye and a thinking brain. Trigonometry gives you the perspective to see such systems.

As I mentioned before, we transition from circular motion to linear motion thanks to the triangle. The relationship between the circle and the triangle is of course made comprehensible through trigonometric functions. Therefore, I think it goes without saying that trigonometry is the place where math and geometry meet the most.

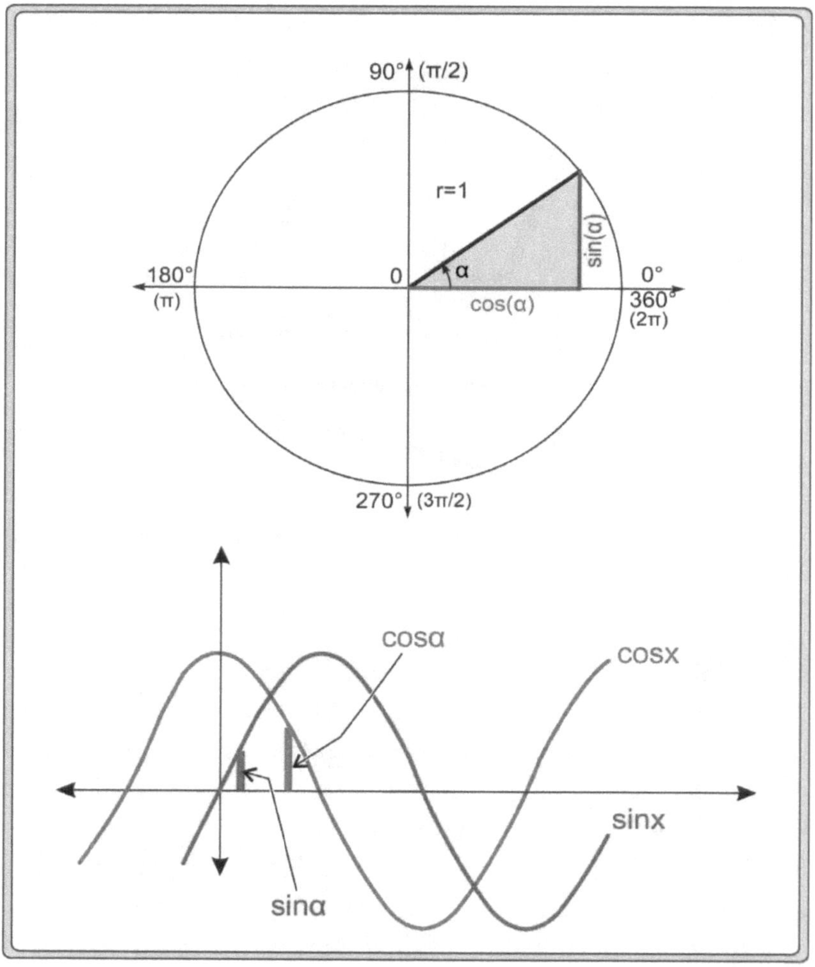

Figure 4.10: In fact, there are always straight lines inside curves

Because we have not internalized the concepts, we hesitate even to express our opinion on any technical topic. Many of us have mixed feelings like, "I wonder if there is something I don't know, what if I am misunderstood, what if what I say is not true?". How do I know? Of course, from myself. When I didn't have a good grasp of mathematics, I remember being very worried about what if there was something I didn't know. When you learn the philosophy of the subject, there is no need to be afraid. When you read this book with this perspective, you will see that you can make much more original analogies and definitions about the relationship between mathematics and the events in life.

The sine and cosine functions, which form the backbone of trigonometry, open the door from linear motion to circular motion or from circular motion to linear motion. Of course, the emergence of these functions is based on the fact that the hypotenuse is oblique. In short, you can say, "The ratio of two lengths opens the door to the curve." When we take the hypotenuse as 1 unit, you can easily see that this ratio corresponds to the length of a side and its projection on the circle is called *an angle*. You can also call it an arc, which is part of the circle. When you draw the graph of this ratio, you can easily see the transition to sine or cosine functions.

The sine and cosine functions are the two basic functions we use to analyze the transformation from direct to curve or from curve to direct. Here I want you to focus on how the sine function is derived. When you see the sine function on the unit circle, that is, on a circle whose hypotenuse is 1 unit, you immediately notice that the sine is actually a line segment corresponding to the length of the opposite side.

When you divide the circle into 180 or 360 pieces, you can see that the circle can be represented by linear, i.e. measurable lengths, rather than curvilinear, which is the basis of analytical thinking. Thanks to this polygon, you will now better understand what it means to analyze life with line segments, the most basic teaching of Linear Algebra.

If you start to see that the arcs on the circle are actually line segments, then you will better understand the expressions with which you can solve a problem. The basic philosophy here lies in the relationship between lines. When you see that what you call the sine or cosine function is actually nothing but a line segment, you begin to solve the mystery of the transition from a static to a dynamic system.

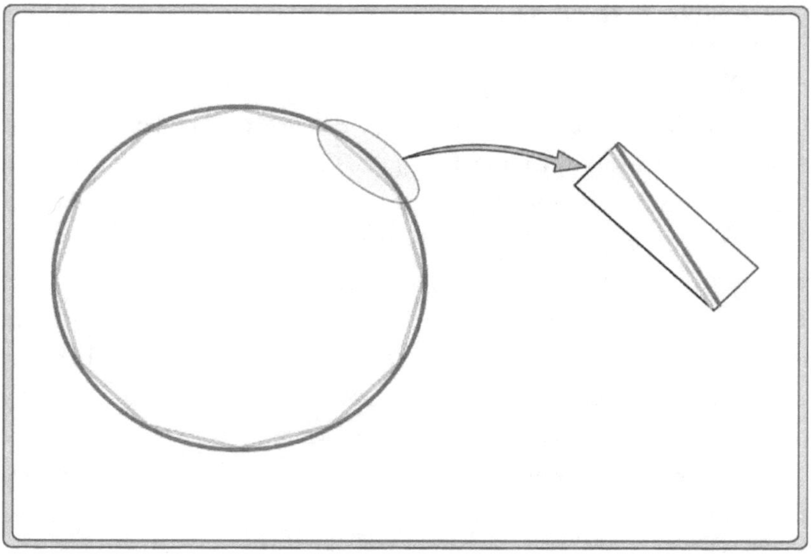

Figure 4.11: Thanks to the lines we see the real picture.

You can see that curved surfaces do not exist when you increase the number of sides of polygons that you can convert into linear lengths. When you increase the number of sides of these polygons to 180,000 or 360,000, you can say without any doubt that the arcs you see from a distance actually correspond to line segments.

If you look carefully at the road lines as you drive down the road, you can see the curvature of the road by these lines. When you get closer to these curves, you realize that they turn into lines. Just as the curves you see when you look too far ahead turn into lines in front of you, you need to see that the curvature you see with sine functions is actually made up of line segments. In summary, the sine and cosine functions are the line segments that make up the curve. Also remember that these functions are derived from the ratio-proportion, that is, by division.

Look carefully at the shape of the sine function, there are always half circles, right? To get that, you divide two-line segments by each other. You could say that the division of two-line segments to get the picture of a circle is the division that has helped human beings the most in solving the mystery of this life.

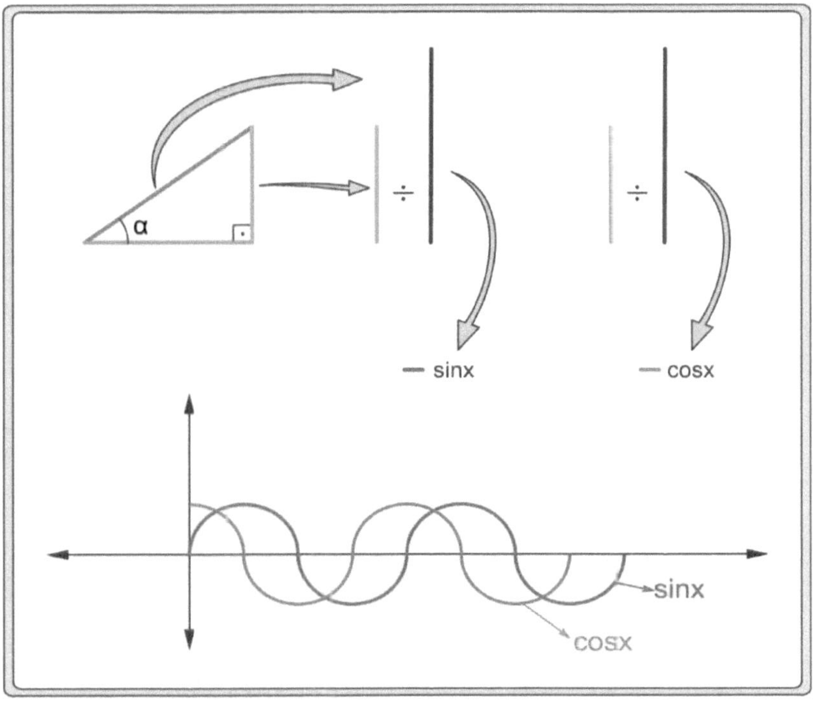

Figure 4.12: See a curve? Then look RIGHT!

The cosine function gives a similar picture to the sine function. Depending on the point of view, the cosine seems to follow the sine a bit behind or in front. One of the great things about these two functions is that no matter what the value of x is, the output of the function has a value between -1 and +1. Since these values don't float away, they are useful in so many places that I can't describe them. You cannot limit tangent and cotangent functions like these. They can go infinitely. The sine and cosine functions, on the other hand, provide very valuable outputs to your problems with controllable and measurable results.

The spirit of circular motion is repetitive motion. In a rotating system, you go around and round and come back to the same place again and again. So, whatever is repeating doesn't necessarily have to be a machine element, a cogwheel or a wave motion. If it is your mental state or your emotions, you can use trigonometry there too. So, trigonometry can be found everywhere, from the formation of seasons to temperature changes, from your mental state to stock market movements.

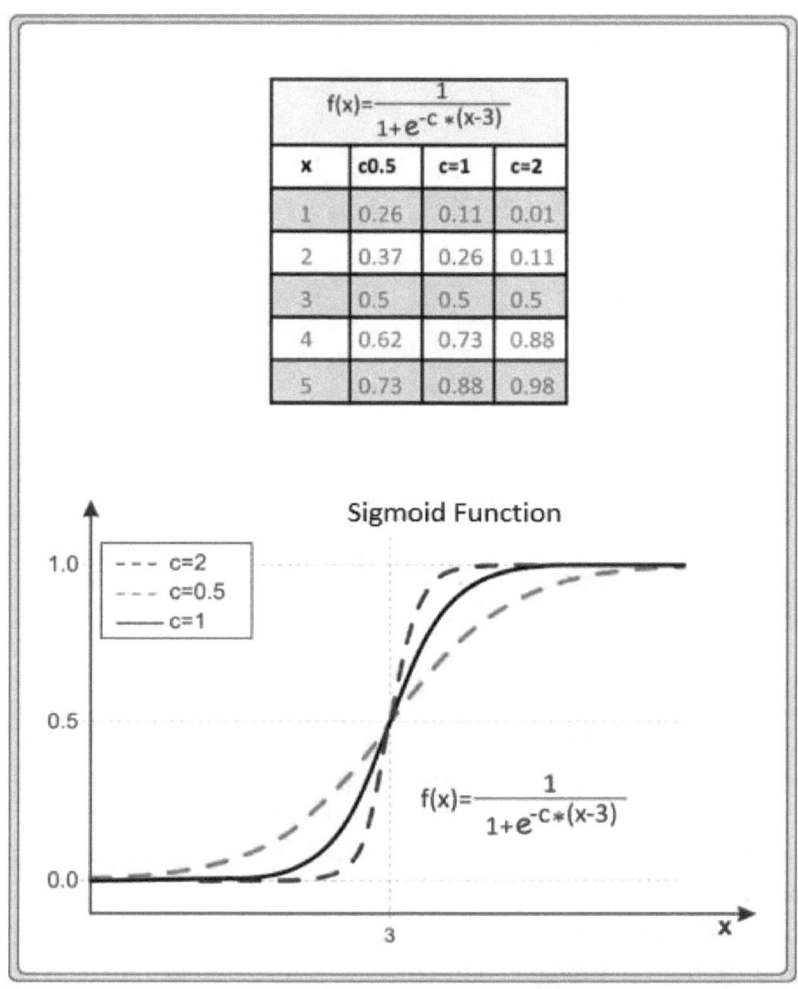

$f(x)=\frac{1}{1+e^{-c*(x-3)}}$			
x	c0.5	c=1	c=2
1	0.26	0.11	0.01
2	0.37	0.26	0.11
3	0.5	0.5	0.5
4	0.62	0.73	0.88
5	0.73	0.88	0.98

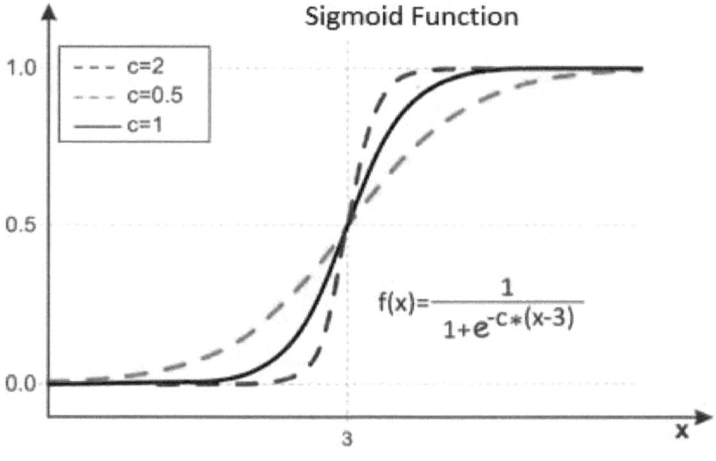

Figure 4.13: There are functions that change the way you look at life.

When I was defending my master's thesis, one of our jury members asked a question "What did you learn in this thesis?". I was using transfer functions a lot in my thesis and I used sigmoid ($\frac{1}{1+e^{-x}}$) transfer function impressed me a lot: No matter what "x" was, the result was a value between 0 and 1. I told the lecturer that this transfer function was the best thing I had learned. I was very happy that no matter what "x" was, the output of the function turned into a value that you could understand, interpret and use. Seeing that the concepts of infinity could be limited made me feel that I could cope with all kinds of problems in this life and I was very impressed by this.

You can tell that the sine and cosine functions, like the sigmoid transfer function, are transfer functions that produce very valuable results when you look at their shapes on the coordinate system. When you look at their graphs, you can see that the outputs of the functions oscillate between (-1, +1). When you multiply these outputs by ten, a hundred or a thousand, you can get concrete tangible numbers. For example, when you multiply the results for c=1 in the previous table by 100, you get 11, 26, 50, 73 and 88. Looking for meaningful expressions to use in your problems? Here, use these.

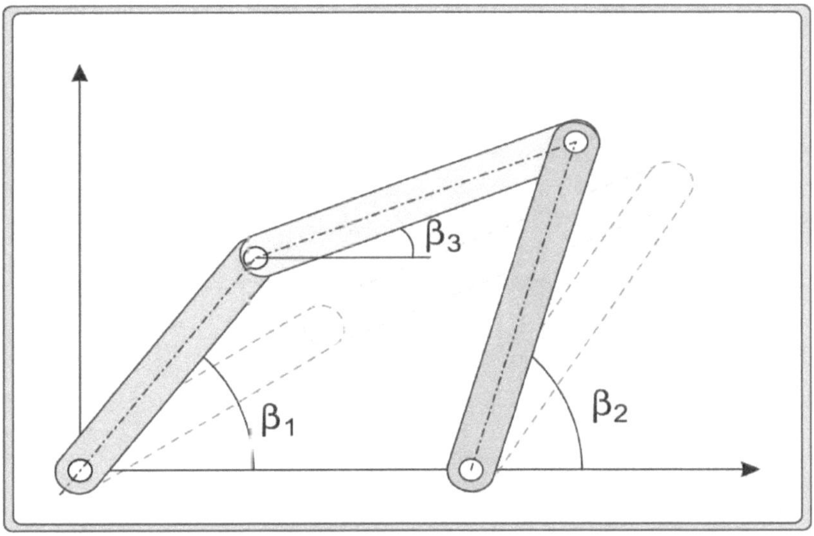

Figure 4.14: Geometry within mechanisms and trigonometry within geometry.

We can say that the age of mechanization was enriched by the mechanisms that emerged by connecting the rods in different ways. You can see these mechanisms everywhere in your life. Examine many systems, from production lines in factories to the propellers and engines of airplanes, even from the landing gear of airplanes to the steering wheels of cars, you can see plenty of such mechanisms. So many mechanisms are produced on the basis of the movement of rods that if you tried to count them, you would lose count. That's why triangle questions are asked to be solved a lot in the education curriculum. When you look at rod mechanisms from this perspective, you will better understand how valuable it is to solve triangle questions.

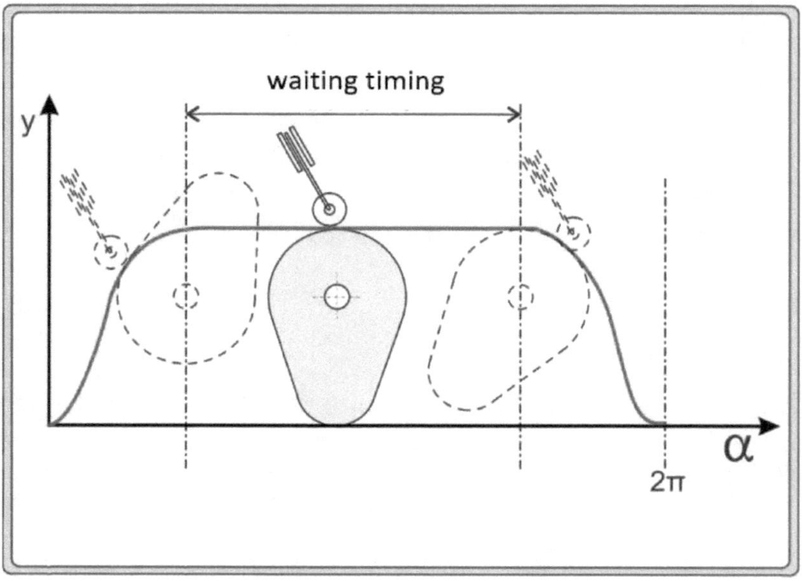

Figure 4.15: Geometry always offers you the beauty to achieve the movement you want.

There is too much circular motion, especially in engines. For example, the cam mechanism used in engines is one of the mechanisms that impresses me the most. I was very impressed by the lids that open and close with the rotation of the egg-like shape around itself. A system that rotates around itself, only allowing the lid to open for a certain period of time, and not allowing the lid to open for a certain period of time, can only be designed so beautifully, right? Is it difficult to make this mechanism? Of course, it is not difficult. When you imagine the movement you want to achieve and put it into mathematics, what could be difficult?

The sine and cosine functions are so important and so many problems have been solved using them that they should be sculpted.

Take a look at the wave function on the next page. As you can see, you can make a picture of this wave using sine functions. No matter how complicated the wave is, no matter how strange it may seem to you, we can make it understandable and measurable by using sine and cosine functions. Now, if I draw you a very complicated graph in the coordinate system and tell you that this graph can be described with sine and cosine functions, would I be saying something wrong?

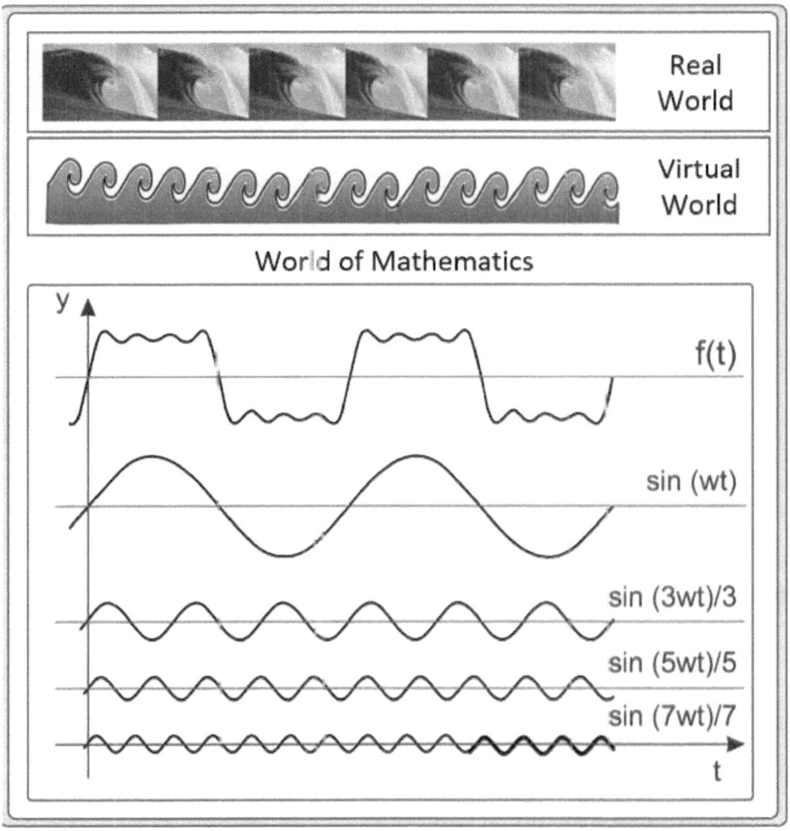

Figure 4.16: Waves are everywhere. Water waves are just one of them.

Take a good look at the shape of the wave motion, what do you see? Shapes similar to the sine function, right? Whenever I am asked a question about waves, I explain everywhere that this subject should be taught as a separate course such as mathematics, physics and chemistry in the last year of high school. The mathematical expressions we use most to understand, interpret and analyze waves are of course sine functions. Isn't the shape of these functions very similar to the shape of wave motion? In fact, concepts such as the circle, right triangle and the number π are taught so that you can understand the sine function. These concepts have no higher task than making sense of circular motion.

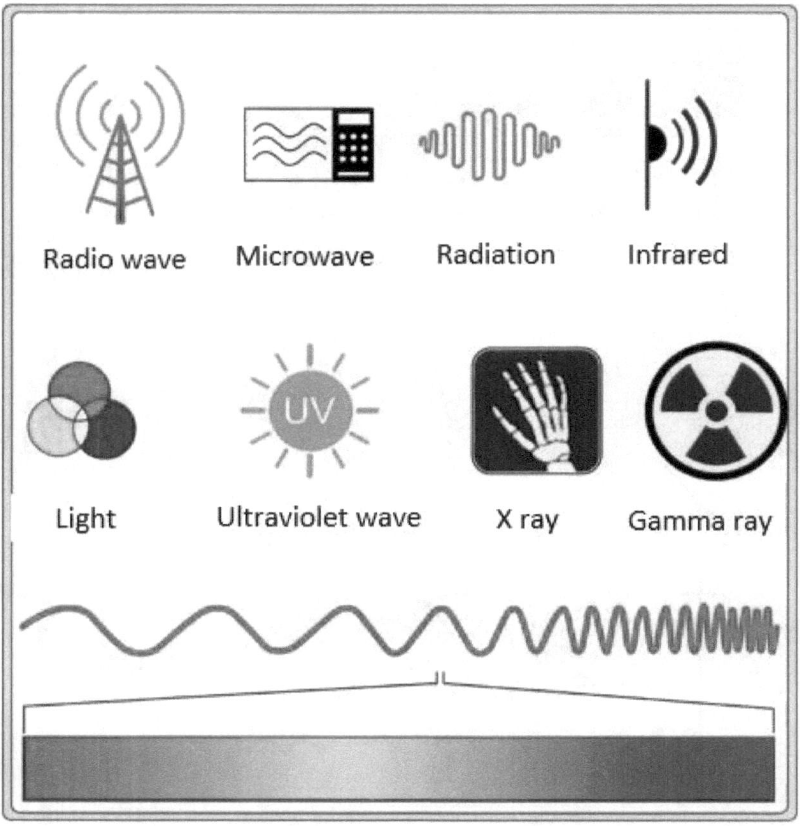

Figure 4.17: Power of the waves

When you learn trigonometric functions, you can understand and interpret the most important topics of scientific life such as light, water, sound, electromagnetic waves, and you can easily create mathematical models of these seemingly complex physical phenomena.

The mathematical model of all events in space, such as the rotation of the Earth around the Sun and the position of the stars, has been transformed into meaningful expressions thanks to trigonometry. From how long it takes the Earth to complete its revolution around the Sun to the position of the Moon relative to the Earth, all kinds of movements in the sky have been solved thanks to trigonometry. It is not for nothing that people in the early ages pondered on these subjects! Because of this curiosity, the history of trigonometry goes back to ancient times. Of course, we can say that the study of space had a great impact on the emergence and enrichment of trigonometry. But as you can see, trigonometry is not only useful in space.

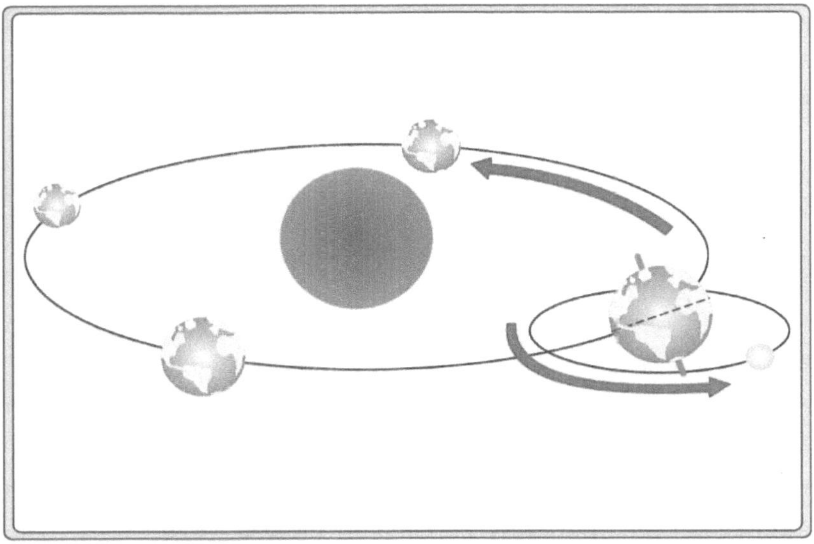

Figure 4.18: The discovery of trigonometry paved the way for our understanding and interpretation of the universe.

Now you may say, "What do we care if the Earth revolves around the Sun, what does it matter if we know this or not, after all, what does it have to do with mathematics?" At first, dealing with celestial bodies such as the sun, stars and the moon seems like a hobby activity with no scientific basis. However, when you look at the developments in the last 300 years, you see that trigonometry has turned into a function that is used not only to solve the motion of celestial bodies, but also to solve problems related to all circular motions in life. Therefore, for us, trigonometry means much more than solving the motion of celestial bodies.

I have explained functions and trigonometry to you. In the following chapters of the book, you will see that derivatives and integrals are also explained from this point of view. "When you internalize the analytical thinking system to solve the problems we face in this life, there is no problem you cannot solve!" If I say that I am not exaggerating, you will understand better when you read the derivative and integral in the following chapters of the book. When you want to solve problems related to electromagnetic waves, the motion of light or atomic energy, which are among the most difficult topics in physics, you will see that your only problem is literature. You can deal with the experimental physics of these topics for years, but you don't need to spend years mastering their literature.

Are you going to solve a problem about electromagnetic waves? You need to know what concepts such as wavelength, magnetic field, electric field, current, frequency, speed of light means. Of course, when you learn these without deviating from the basic philosophy of analytical thinking, it will not be difficult for you to master the literature of electromagnetic waves. If you can transfer the functional relationships you need in the problem to the coordinate system, mathematics will do the rest. You don't need to study for years to solve problems related to electromagnetic waves. If you know where mathematical expressions appear, you can master this literature in 2-3 months. Of course, this period will not be longer when you internalize mathematics and gain a scientific perspective.

If you have a good analytical perspective, you can, of course, get a bachelor's degree in chemistry and do a PhD in economics, or a bachelor's degree in aeronautical engineering and do a PhD in electronics engineering. Don't forget that mathematics is behind the fact that you can do master's or doctoral programs in different fields after your undergraduate education!

Logarithm

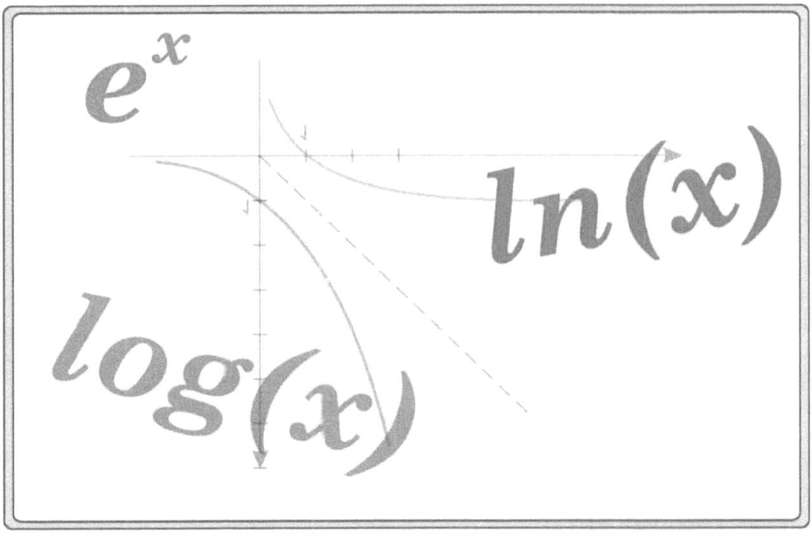

Why Do We Take Logarithms?

I would like to tell you about logarithm, which is one of the most interesting subjects to make people memorize. When I was in high school, I was told that when the British ruled India, they made Indians memorize the logarithm ruler to prevent them from using their brains too much. Of course, I didn't memorize the logarithm because I didn't want to fill my brain with empty things, but I thought I would make an airplane by memorizing its rules. Years later, I realized that there was no point in that, because I had learned logarithms without knowing where and how to use them in real problems such as integrals and derivatives.

I constantly emphasize in various parts of the book that you need to know where and how mathematics is used. In fact, this sentence summarizes the main idea of the book. Of course, you have encountered people who think and say that they know where and how mathematics is used. Pay attention to the common characteristics of such people. Almost all of them are like students who have memorized their lessons, and the answers they give are almost exactly the same. Because as they memorize the rules, they also memorize where they are used.

When I researched how logarithms are explained in the current education system, I came across familiar sentences in those lines while reading an article about its use in space exploration. Just like with functions, they bring the subject to space and prevent the understanding of the issue. We need mathematics in this world. When will we give up this kind of understanding that takes mathematical concepts into space and prevents us from perceiving the spirit of life on earth!

If you search carefully, you can find Turkish texts explaining where the logarithm is used. For example, when you ask the question "Where is the logarithm used?", almost all sources immediately tell you that it is used to measure the intensity of sound, to find the PH value of acid and base, as if the logarithm is not used anywhere else. When you ask, "Are there any other examples?" the texts fall silent and leave you hanging like a nightingale. These are all trite memorized answers translated from English, so these answers have no soul. Those who are good at English add the intensity of the earthquake to this. If the logarithm is used for the intensity of sound, the intensity of the earthquake and the PH, don't you think it's better not to learn the logarithm at all?

Those who take a more general approach to the subject explain that logarithms make it easier to express large numbers. I don't care about numbers! I want to be told about the usefulness of mathematics in understanding and interpreting this life and how this happens. By the way, I also like dealing with numbers, so why bother with logarithms, right?

If you don't know what logarithms are for, you might think that if you memorize the logarithm ruler you will build an airplane. So, when a higher mind says, "When you memorize logarithms, you will build an airplane," of course you want to memorize it.

You can memorize poetry, you can memorize the text of a speech, you can memorize the lines of an actor, you can also memorize the rules of mathematics. I don't want to say that memorization is completely bad. We just have to learn where the mathematical concepts you memorize come from and what they do. Otherwise that memorization is useless.

Figure 5.1: You cannot learn logarithms by memorizing them.

If you want to make scientific thinking dominant, what you need to do is not to memorize, but to explain where it comes from and what it can do.

Logarithm, like the other rules of mathematics, emerged to solve the problems encountered as a matter of course. People's curiosity to study celestial bodies from the earliest ages turned into curiosity to study very, very small structures such as cells and atoms in the 16th century. I can say that the problem of not being able to show the magnitudes of very large and very small structures in the same picture that emerged as scientists navigated in very large and very small structures became more apparent and this triggered the birth of the logarithm.

When you look at its mathematical meaning, it is explained that logarithm is a mathematical function developed to see how arithmetic and geometric sequences change with respect to each other. Napier, the inventor of logarithms, is said to have invented logarithms to describe the relationship between arithmetic and geometric sequences.

You may have noticed the similarity between the arithmetic sequence 1, 2, 3, 4, 5 and the geometric sequence r^1 , r , r , r^{234} , r^5 . Now, let's say we have 2 tools. One of these vehicles travels as in the arithmetic sequence while the other travels as in the geometric sequence. For example, the first vehicle travels as 1, 2, 3, 4, 5 while the other vehicle travels as 3, 9, 27, 81, 243. You can of course show this as 3^1, 3^2, 3^3, 3^4, 3^5. As you can see here, I chose r as 3, of course you can also choose r as 10 or 20. Now, if we were to draw the graphs by directly processing the real values of these numbers, that is, if one number line increased as 1, 2, 3... while the other number line increased as 3, 9, 27... then you would see that it is almost impossible to control this graph, right? We can say that the logarithm is actually hidden inside these number lines. Logarithm is the name of the meeting of these two number lines in a meaningful place!

As I mentioned before, let's expand on this a bit; as you know, we are mostly looking for solutions to physical problems by transferring them to the coordinate system. In a coordinate system, each axis is a number line. To understand this, let's transfer the above quantities to the coordinate system. Now imagine two number lines on a plane as they are, that is, without logarithms, do you think it is easy to deal with numbers with one or two digits on one side and numbers with 40-50 digits on the other side and to bring these two number lines together on the same plane?

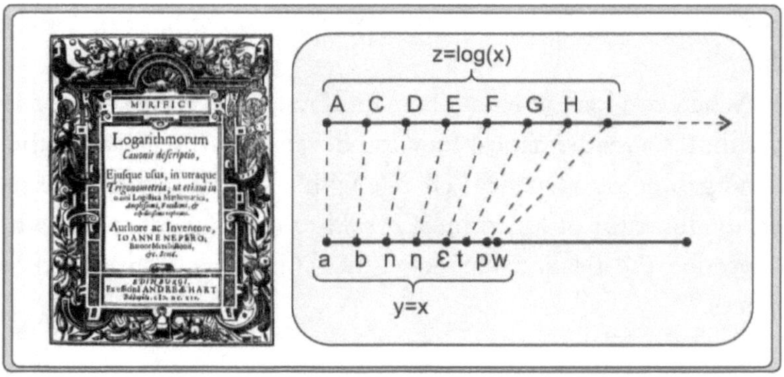

Figure 5.2: Napier's Logarithm.

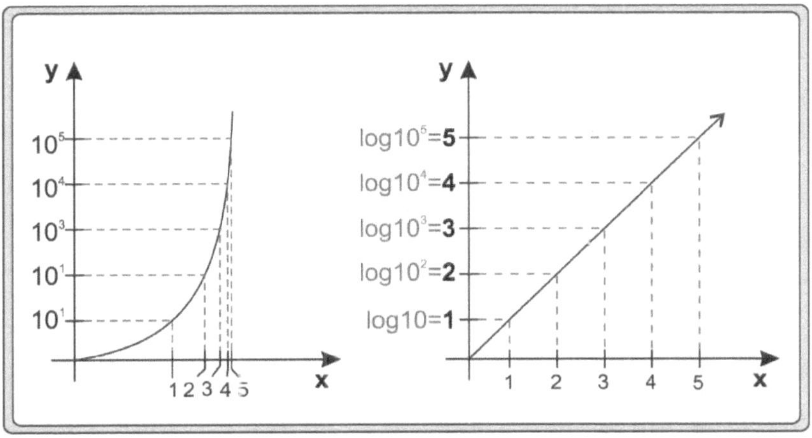

Figure 5.3: Logarithms also linearize curves.

Let's show the relationship between these two sequences on a graph and see what we can do to get meaningful expressions. In the logarithmic representation here; on the x-axis, you will see the numbers increasing in the form of arithmetic sequences, i.e. numbers increasing in a uniform way such as 1, 2, 3, 4, 5, and on the y-axis, you will see the logarithm of the numbers increasing geometrically.

There are plenty of such relationships in daily life. You may say, "Where is it, I can't see it?" Of course, if we did a lot of experimental studies, we could easily see them too. But unfortunately, we deal very little with experimental data. In fact, we take them as they are and use them. If you say, "I have never seen the logarithm," I would say, "Take a look at the world of measurements." Look at how transitions between quantities are made with logarithmized numbers, how meters and grams change, and you will witness how problems are defined and solved with quantities whose logarithm is based on base 10.

We don't always solve problems where kilograms and meters are together. Sometimes it is necessary to define and solve many problems where grams and meters or kilograms and millimeters meet in the same plane. Then you will see more clearly that solutions are sought with logarithmized quantities.

There are several unit systems in the world; human beings invented and developed them. At the end of the day, unit systems are definitions, and we use them to try to make sense of the graphs we draw. Because of the nature of the description, you can find many cases where one side increases smoothly and the other side changes abnormally because of the pictures that the unit systems produce. Of course, with the unit systems you use, you get a picture and you solve a problem. If we were to use another unit system, then we would get a completely different picture, and according to that picture, we would main lize and interpret things differently.

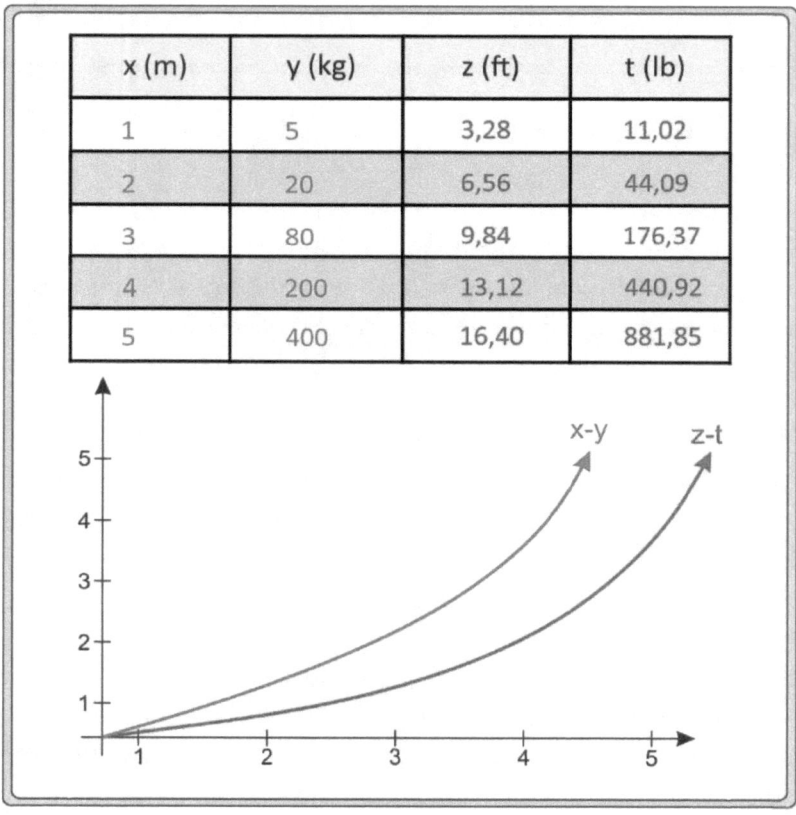

x (m)	y (kg)	z (ft)	t (lb)
1	5	3,28	11,02
2	20	6,56	44,09
3	80	9,84	176,37
4	200	13,12	440,92
5	400	16,40	881,85

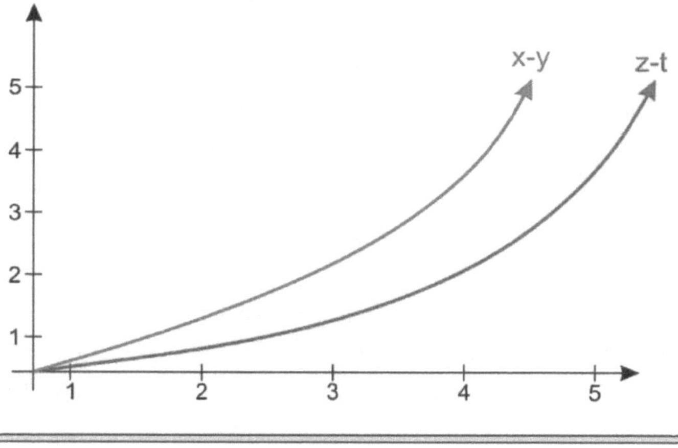

Figure 5.4: Different graphs are obtained with different unit systems.

Pay attention to the graphs above, an experiment with the same result can be graphed very differently using two different unit systems.

Now let me try to explain what logarithms do with a simple example:

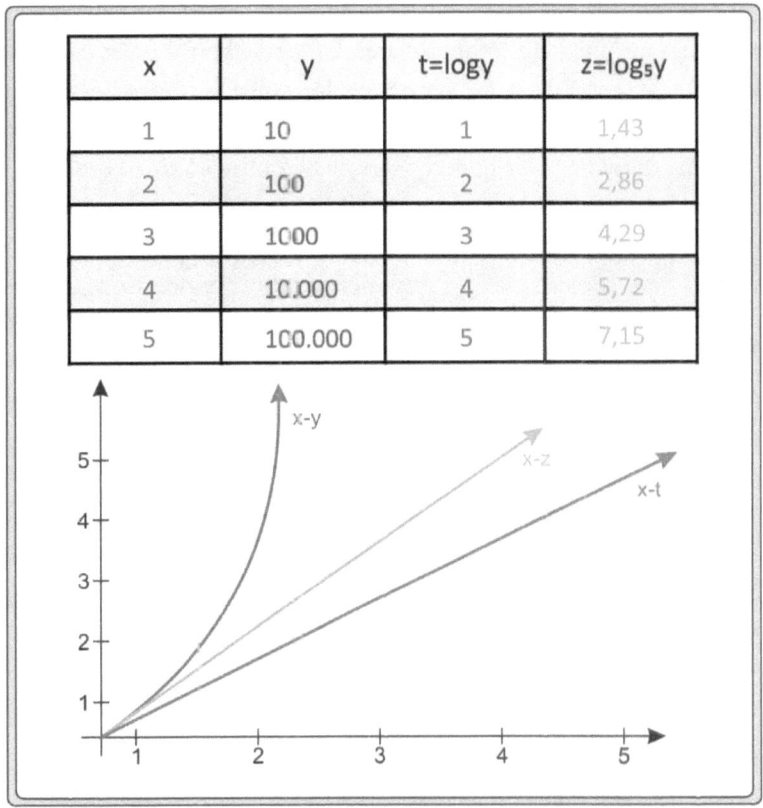

x	y	t=logy	$z=\log_5 y$
1	10	1	1,43
2	100	2	2,86
3	1000	3	4,29
4	10.000	4	5,72
5	100.000	5	7,15

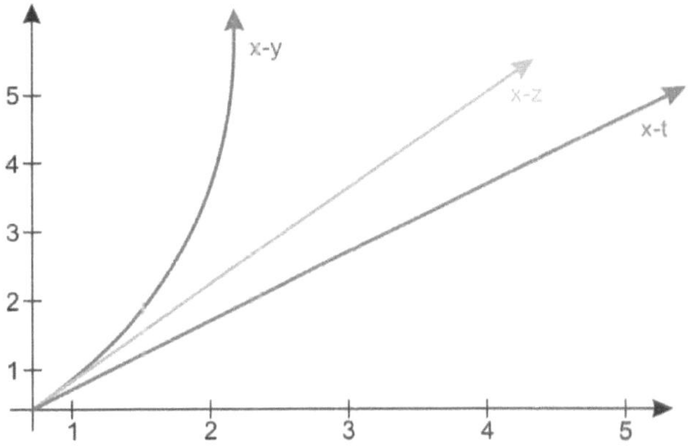

Figure 5.5: Inside the logarithm is the spirit of Linear Algebra

Now draw the x-y, x-z and x-t graphs, which do you think gives clearer and more understandable information; the x-t graph, right? Look carefully at the table, you will see that both axes are composed of integers. In the spirit of Linear Algebra, the graph is composed of lines. You won't get confused when you want to read a graph made of lines. The x-z graph is not bad either, but the x-y graph looks like a picture of uncertainty. Linear Algebra is actually the art of understanding life with lines, right? Do your operations with logarithmic quantities that will give you the most beautiful lines, try to get a clearer picture there. Don't wait to be told why you should take logarithms according to logarithm 8 or logarithm 15. Choose the base that gives you the clearest picture.

You may face serious difficulties in understanding and interpreting curves. There are plenty of curves in daily life. Curves represent irrationality and lines represent rationality. As you know, the set of irrational numbers has many more elements than the set of rational numbers. Irrational, as the name implies, means unpredictable, infinite.

This universe is designed irrationally in the spirit of eternity, but we rationalize and make sense of all the events in the universe. To understand this, one day I picked up a meter and measured my height. I was 186 cm tall. If I had measured more precisely, the result would have been 1.864 mm. If I had measured more precisely, it might have been 18.584 µm. As you can see, even the measurement of your height is heading towards irrationality. Don't these facts make it seem like the universe is irrationally constructed and we are being asked to try to solve the mystery of this universe? This is also the reason why the speed of light is known to be 300,000 km/s. If you cannot make sense of all this in rational terms, then you cannot get out of it. In summary, I can say that you cannot solve any problem that you cannot rationalize.

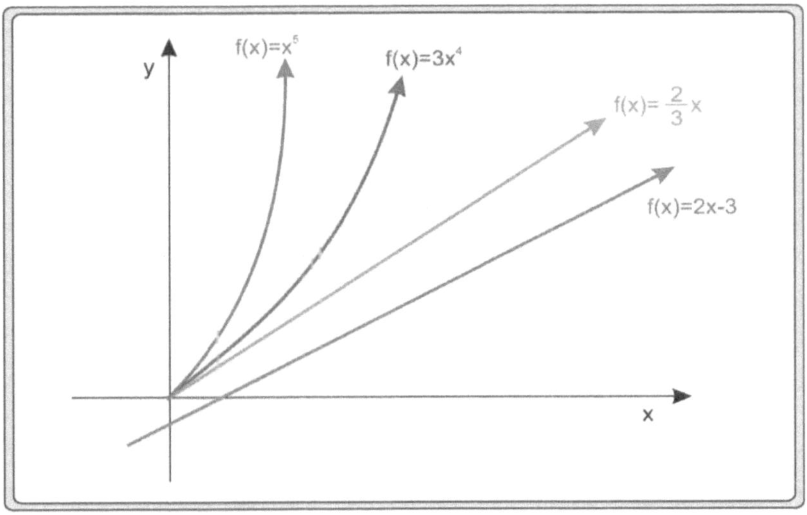

Figure 5.6: You understand and solve problems when you linearize them.

For example, when you draw the curve $f(x) = x^5$, the resulting shape may not make much sense to you. Yes, we have difficulty perceiving geometrically increasing sequences of numbers. We are always more comfortable understanding and interpreting an arithmetically increasing sequence of numbers. Geometrically increasing numbers involve multiplication; the brain has a very difficult time perceiving multiplication. But arithmetically increasing numbers involve addition, which is not so difficult to perceive. Geometrically increasing numbers open the door to curves. Arithmetically increasing numbers tell us the truth.

Since the graphs of functions such as $f(x) = 3x^4$ or $f(x) = 5x^6$, $f(x) = 2x^7$ look like curves, you will have difficulty interpreting them. But since the graphs of functions such as $f(x) = 3x + 6$, $f(x) = 2x + 5$, $f(x) = 5x$ consist of lines, it will be easier to perceive and interpret them. Here, logarithmic functions provide you with expressions that allow you to make the transition from curve to line and in this way, you can better understand and interpret the problem.

It is clear that a new concept is needed to draw meaningful conclusions from coordinate system graphs of both celestial bodies and small structures like atoms. The reason why I am explaining logarithms in terms of celestial bodies and atoms is to make you understand the difficulty of dealing with very large numbers in one case and very small numbers in the other, and the difficulty of representing them in the same picture.

In any experimental result you can get such very large and very small data. You may find it difficult to make sense out of them. We can say that this difficulty is overcome by logarithms. The only purpose of mathematics is to get results that are meaningful for you, so I guess logarithms can't have any other purpose, what other purpose can they have!

For example, let's take a look at what we know about the weights of small and large particles! The Planck particle is the smallest substance found so far. The weight of this particle is calculated to be about 0.000.000.000.000.000.000.000.000.000.000.000.0001 grams. Let's choose the Sun as a large object. The weight of the Sun is calculated to be about 1,000,000,000,000,000,000,000,000,000,000,000 grams. Now write down these quantities. It doesn't seem very easy to operate with these representations of these numbers, does it?

I couldn't even count the zeros. Adding, subtracting, multiplying and dividing these numbers is not impossible. Now let's represent them as exponents: We can show the weight of the Planck particle as 10^{-34} grams and the weight of the Sun as 10^{30} grams, which makes our work easier. What about the weight of an object 10 times bigger than the Sun? Just increase the exponent from 30 to 31. In other words, doesn't it look nice to express that the real weight increases 10 times when we increase the exponent by one? But life is not so simple that it can only be explained with exponents. There are many numbers between exponents, and remember that you can only navigate between these numbers with logarithms!

Here we need to open parentheses for exponents and radicals, arithmetic and geometric sequences. I want to emphasize that exponents and radicals, arithmetic and geometric sequences are taught to help you internalize logarithms. Of course, these concepts are also used in conjunction with other mathematical functions, but one of the most valuable purposes of learning exponents and radicals, arithmetic and geometric sequences is to help you better understand logarithms. Geometric sequences are multiplicative and this is where exponents come from. This is where exponents and geometric sequences meet the most.

If you do a research on "Why are these things taught in the

education system?", believe me, you will return home without finding the answer. You may refer to logarithms in some texts, but the aspect of logarithms that touches life remains incomplete. Now, when you learn the aspect of logarithms that touches life, of course, you will also grasp the logic behind teaching exponents and roots, arithmetic and geometric sequences, and you will be happy to learn these concepts. These concepts carry you to logarithms and logarithms carry you to life.

It is worth reminding again that the mathematical concepts taught in all primary education periods, from ratio and proportion to exponents and radicals, are taught to make sense of the most valuable subjects of mathematics such as trigonometry, logarithm, derivative and integral, which help to establish a connection with this life!

As you know, logarithm is defined as the mathematical function that is the inverse function of exponential expressions and is shown as $\log_a y = x$ if $y = a^x$. Thus, the logarithm of the number 10^{30} in base ten is 30 and using this number will give you more meaningful results when necessary.

Logarithm is actually an abbreviation! Logarithms make your operations much simpler to see. If there were no logarithm, you would get lost in numbers. As you can see, the logarithm is a mathematical function that allows you to navigate between very small and very large structures. Thanks to the logarithm, you can combine the distance between the galaxies in the universe and the distance between the particles of an atom in the same square. In this way, you will not get lost among real numbers. With logarithms, you can transform all meaningless graphs into meaningful ones.

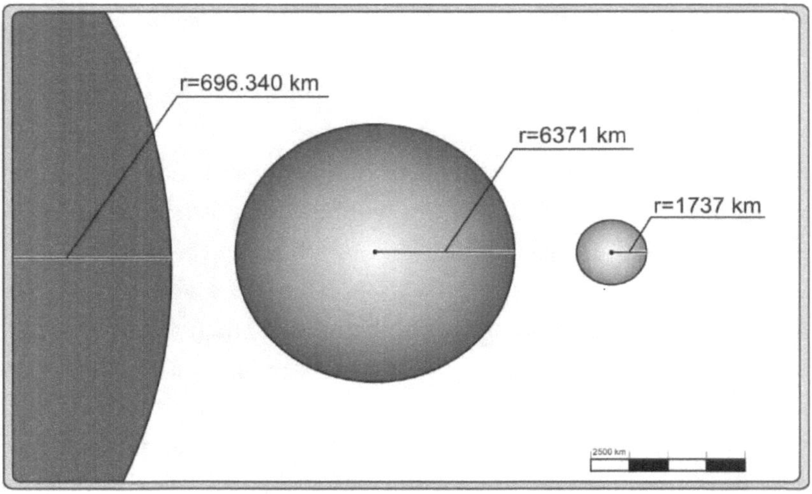

Figure 5.7: In reality you cannot fit the Sun, the Earth and the Moon into the same picture.

Now let's continue explaining the subject through a simple application. We will make a picture with the Sun, the Earth and the Moon, staying true to their measurements. If we give their measured sizes proportionally: If the diameter of the Moon is 10, you can consider the Earth as 40 and the Sun as 4400. Now, if you try to show these three in the same picture using the coefficients, if you show the Moon with a circle with a diameter of 5 cm, you have to show the Earth with a diameter of 20 cm and the Sun with a circle with a diameter of 2200 cm (only the Sun has a diameter of 22 m).

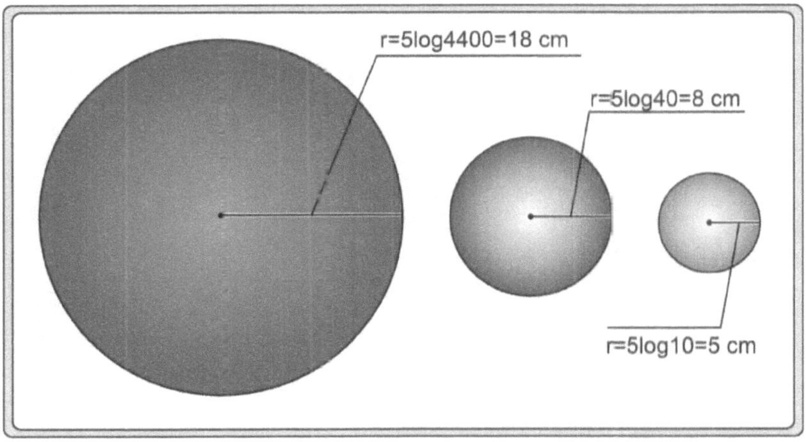

Figure 5.8: Sun, Earth and Moon in the same picture thanks to logarithm.

Using the actual measurements, can you fit them on an A3 sheet of 29.7 x 42.0 cm paper? If you look carefully at the radii, you will immediately realize that they will not fit.

Now let's take their logarithm to base 10. We get log10 = 1, log40 = 1.6 and log4400 = 3.6. Using these coefficients, if you show the Moon with a circle with a diameter of 5 cm, you can show the Earth with a circle with a diameter of 8 cm and the Sun with a circle with a diameter of 18 cm, and you can easily fit them on A3 paper.

Country	Population	Log(Pop.)	Land	Log(Land)
Andora	77.000	4.9	47	1.67
Tunus	12.000.000	7.07	164.000	5.21
Turkiye	82.000.000	7.9	780.000	5.9
China	1.400.000.000	9.15	9.600.000	7

*from open sources

Figure 5.9: Logarithm keeps you in touch with reality

Now imagine the population or the area of Andorra, Tunisia, Turkey and China on a graph. I am not going to draw their graphs, just imagine and try to see the reality. If you transfer the current situation to the coordinate system as it is, you will see that you get complex shapes, just like the Earth, the Sun and the Moon. But if you take the logarithm of the current situation and graph it, you get shapes that make more sense. The logarithm is a mathematical tool that provides a unity of meaning between quantities without breaking the connection with the real situation, giving us the opportunity to see the big picture more clearly. Logarithms help you not to get lost in the details.

But it cannot be done with logarithms alone. The logarithm gives you an approximation, but you have to make your own picture. With the logarithm you find an approximation, but it is your interpretation that will make it look like reality. Play with the numbers you find with the logarithm like a ball and get the most visually beautiful proportional relationship and draw the graph. It's all as simple as that.

As you know, graphs are actually another name for mathematical painting. Scientific developments have evolved by analyzing and interpreting these pictures. Of course, these pictures you draw need to tell you something. That's why a good mathematician or engineer needs to draw and interpret them correctly. Look at the graphs on the next page, drawn with different magnitudes, and decide which one makes more sense. If you can't make a harmony between the graphs, you won't be able to see any difference between the pictures.

x	y_1	y_2	log y_1	log y_2
1	5	8	0,70	0,90
2	60	250	1,78	2,40
3	725	120	2,86	2,08
4	12.000	400	4,08	2,60
5	270.800	1.000	5,43	3,00
6	252.000	5.000	5,40	3,70
7	658.000	10.000	5,82	4,00
8	1.254.000	1.300.000	6,10	6,11
9	5.450.000	5.600.000	6,74	6,75
10	15.000 000	27.500.000	7,18	7,44
11	658.000.000	845.000.000	8,82	8,93
12	1.580.000.000	1.780.000.000	9,20	9,25

Figure 5.10: Example data table showing the difference of logarithm

You can consider the values in the table as the results of two different experiments. Now, if you take the values of y_1 and y_2 as they are and graph them, you will get almost the same graph. Someone looking at these graphs might think that the results are the same. If you take the logarithms of y_1 and y_2 and graph them, then you will get a more realistic and interpretable graph.

I deliberately gave the values in the table without using units. You can see them everywhere. From the expansion of gases to the tensile strength values of metals under different temperatures, it is possible to obtain very, very different results in many problems.

When you look around the world, we can see dozens of countries with populations ranging from countries with three-digit numbers to countries with 9-10-digit populations. It is also possible to talk about many different product groups produced by these countries, from agricultural products to industrial products. These values can of course be expressed in many, many different numbers, so logarithms will help you when you want to show and analyze them in the same graph.

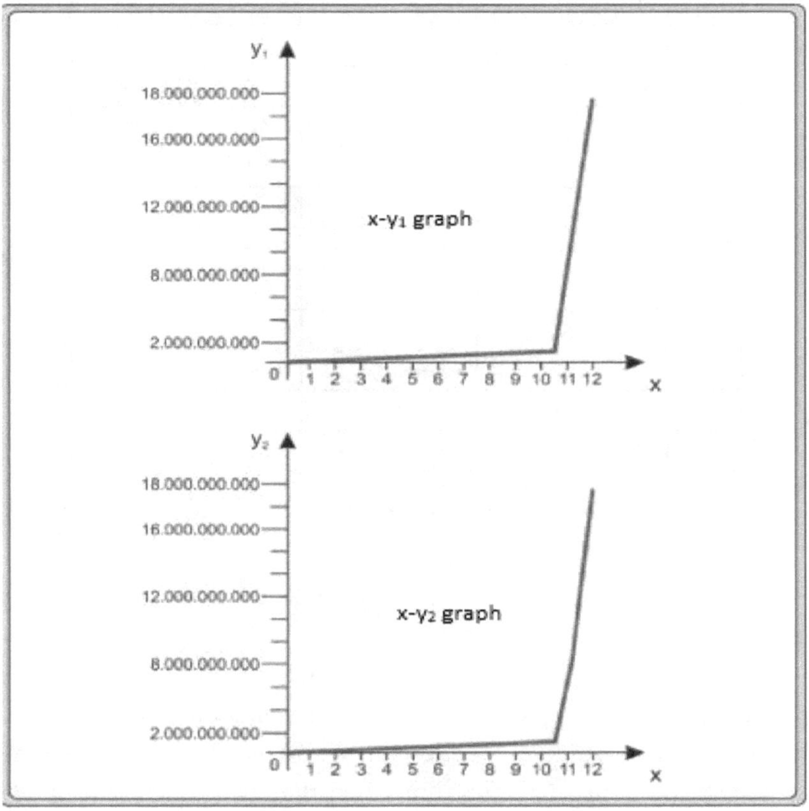

Figure 5.11: Example graphs showing the difference of logarithms.

Pay particular attention to empirical data and don't get lost navigating through them. You will not get lost if you use logarithms. Since we don't deal with a lot of experimental data and we simplify the experiment to finding our own weight, it's normal that logarithms are not a necessity for societies like ours. But there are thousands of them in everyday life, just like undiscovered mines, just waiting to be found.

Graphs drawn without logarithms all look alike. Now how do we distinguish them, which one tells us the correct results? Can anyone understand this without taking the logarithm? The logarithm allows us to get meaningful expressions from these graphs. Logarithm is a necessity and it offers you such beauties. Thanks to logarithms, you can read and interpret the difference between them much better.

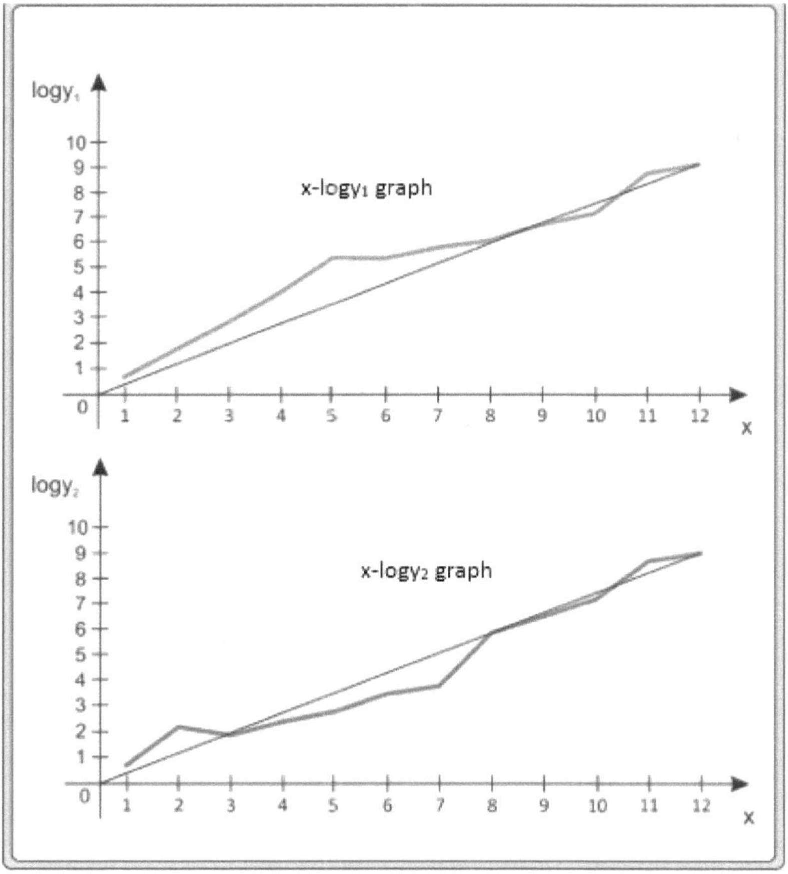

Figure 5.12: The logarithm shows you the unseen truth.

As you can see from the graphs, logarithm is a mathematical operator that was created to make sense of the results. If you have a data set and you want to make sense of it, you may need to use logarithms. Mathematics is a language that has emerged to understand this universe and this language is enriched with logarithms.

Now, when talking about logarithms, it would be impossible not to talk about the number e. You cannot decipher mathematics without understanding e, the most valuable number in mathematics after π. Let's try to understand what this number tells us, how it came to be and what it does.

Special Section
Number "e"

$$e^{i\pi}+1=0$$

$$e = \lim_{n\to\infty}\left(1+\frac{1}{n}\right)^{n}$$

$$e = 2.71828818.....$$

The Number "e", Another Name for Multiplication Together

The logarithm taken with the number e is known as *the natural logarithm.* According to this definition, you can probably call the logarithm taken with other numbers *an artificial logarithm.* When I think of the number e, I immediately think of the number π. e is like the little brother of π. Why is the logarithm based on the number e so important? When you look at the logarithm, for example, why should we deal with the logarithm of an irrational number like 2,71828..., which is known as the number "e", that is, a number that goes to infinity, when we can take the logarithm with smooth numbers like 2, 3, 5 and 10? Why is this number so important and why does it occupy so much space in the education system?

When you look at its history, there are many different stories told with the number e. It is said that John Napier, the inventor of logarithms, referred to this number, Jakob Bernoulli used it, and the legendary mathematician Leonhard Euler gave it his name. The number π has a counterpart in our world and is very easy to perceive. We all know that when we divide the circumference of a circle by its diameter, the irrational number starting with 3.14 and going to infinity is called the number π. But when it comes to the number e, no one remembers anything.

Of course, not only in our country, but also in the world, the number e is not as popular as the number π. We have a picture of the number π in our minds with the concepts of circle and diameter. As you know, March 14 is a special day for the number π, but the number e does not even have a special day.

The number e is said to have emerged with the increasing relationship between money and interest with the beginning of the industrialization period. It is rumored that the number e was found as an answer to the question of what the amount of money at the end of the year would be if the simple interest taken as a year was taken monthly or daily and these were added to the principal, which is called compound interest. I don't know whether this information is absolutely true or not, but I can tell you that you can get the number e by doing this process.

Since my first acquaintance with the number e was through the concept of interest, I thought that this number was learned for banking and economics education. When I encountered problems related to the number e in my university years, I could not see the truth behind it and could not establish a connection with interest, and I was content with just solving the given questions. However, years later, I can say that I realized that interest problems were just a tool to understand and comprehend the number e.

In order to better understand how the number e came about and what it wants to tell us, let us first take a look at what simple and compound interest are.

Simple interest is the first thing that comes to mind when we think of interest. With simple interest, you have one principal and when you invest this money at 100% interest per year, your money doubles at the end of the year. With compound interest, this period is divided into periods and at the end of each period, your principal enters the new period as principal along with the interest you received in the previous period. In other words, the interest you receive immediately turns into principal and starts to multiply together. While simple interest deals with the initial form of your money, compound interest deals with the final form.

Let us continue explaining the subject by formulating interest. Let's call the principal you have a and the interest received at the end of a year p. Let's call the total money you will have at the end of the years. Of course, the total money at the end of the year is equal to the sum of principal and interest, which we can denote by s = a + p.

The interest rate is the ratio of interest to principal. We can represent this as n = p / a. If you define the interest rate as a percentage, you multiply n by 100. For example, if p/a = 1, interest rate 1 is also 100%. If p/a = 1/2, your interest rate will of course be 50%.

I would also like to remind you that the duration of the interest is proportionally related to the duration you set, rather than days, months or years. Statements such as "I received 500 liras or 2500 liras in interest!" are meaningless on their own. You need to know how long the interest you have received and establish a proportional relationship between them. You find this proportion by the ratio of the time remaining in interest to the total time.

Let's assume that you receive 100% interest per year. The amount of interest we will receive at the end of the year $p = a \bullet n \bullet t$ is calculated with the equation. Here, since the interest rate n is 100% and t is one year, both are equal to 1. $p = a \bullet 1 \bullet 1$ from the equation $p = a$ you'll find it.

Your total funds $s = a + p$ s = a + a = 2a. If your money is 100 liras, the total amount of money you have after one year at simple interest will be 200 liras. Compound interest is perceived slightly differently from simple interest because it has a term concept, but in fact interest is interest.

Here, if you divide the annual interest into periods, the interest rate you will receive will be the ratio of the period in that period to the year. For example, if you invested money in simple interest to receive 100% interest for 12 months and you have 3 months left, the amount of interest you will receive will be 100 * 3/12 = 25 liras. I think you can see more clearly from this example that the amount of interest is proportionally related to both the interest rate and the duration.

Within the framework of the logic I have explained, if you have 1000 liras in your hand and you have set an interest rate and a time period that you will receive 400 liras of interest after 800 days and you need to find the money you will receive after 200 days, all you need to do is to divide and multiply the numbers by each other.

Let's try to explain how to reach the number e after all these proportional relationships I mentioned through a simple example. Let's assume that you have 100 lire and you deposit this money in a bank that pays 100% interest per year.

At simple interest, this money will be available after one year. n ve t $1'$eşit olduğu için $p = a \bullet n \bullet t$ equals the interest to the principal ($p = a$) you can calculate the total amount of money you have as 100 + 100 = 200 liras using the equation s = a + p.

Now let us divide the year into 4 periods and see what the result will be;

The coefficient for the time in interest, t, will be equal to 1/4 since there are 4 periods in this example. The interest rate n is equal to 1 due to 100% interest.

$s = a + p, \; p = a * n * t$

When you combine these;

$$s = a + a * n * t$$

When you write the equality and then put it in common parentheses;

$s = a(1 + n * t)$ you can write.

$n = 1, \; t = 1/4$ when you put the values in place;

$$\boxed{s = 100(1 + 1 * 1/4)}$$

You will reach the result.

At the end of the first semester, the amount of money you will receive together with your principal is$_1$ = 100(1 + 1/4) and 100 + 25 = 125 liras.

Since the interest in the second period is added to the principal at the end of the first period, you now have a new *principal* of 125 liras.

At the end of the second semester, the total amount of money you have will be s = s_{21} (1 + 1*1/4) and s_2 = 125 (1 + 1/4) and 125 + 31,25 = 156,25 TL.

If you had kept your money in simple interest for the first two periods $p = a * n * t$ When you substitute n and t in equation 100*1*2/4, you get 50 liras of interest, whereas here you get an additional 6.25 liras of interest. This money is the proportional return in the second period of the interest of 25 liras from the first period.

In the third period, we find 156.25(1 + 1/4) = 195.3125 with the equation $s_3 = s_2$ (1 +1*1/4) with s_2 being the residual principal.

This money is defined as new principal at the beginning of the fourth period and at the end of the fourth period; $s_4 = s_3$ (1 +1*1/4); s_4 = 195,3125(1 + 1/4) = 244,015625.

As you can see from these equations, both the principal and the interest earn interest together for as long as they remain in the bank, increasing the amount of your money.

In compound interest, the total money is the product of the principal of each period times the total return of the previous period, which can be written as follows

$S =_1 a(1 + n * t)$

$S = S_{21}(1 + n * t)$

$S_3 = S_2(1 + n * t)$

$S_4 = S_3(1 + n * t)$

The money you will receive at the end of the fourth semester;

$s_4 = s_3 * s_2 * s_1$ We can show it with equality. When you substitute the expressions;

$$s_4 = a(1 + n * t)(1 - n * t)(1 + n * t)(1 + n * t)$$

When you substitute n and t;

$$s_4 = a(1 + 1 * 1/4)(1 + 1 * 1/4)(1 + 1 * 1/4)(1 + 1 * 1/4)$$

$s_4 = a(1 + 1/4)^4$ you can write.

Also, from here;

$s_4 = 100(1 + 1/4)^4$ You can calculate =244,015625 directly without finding the result for each period separately.

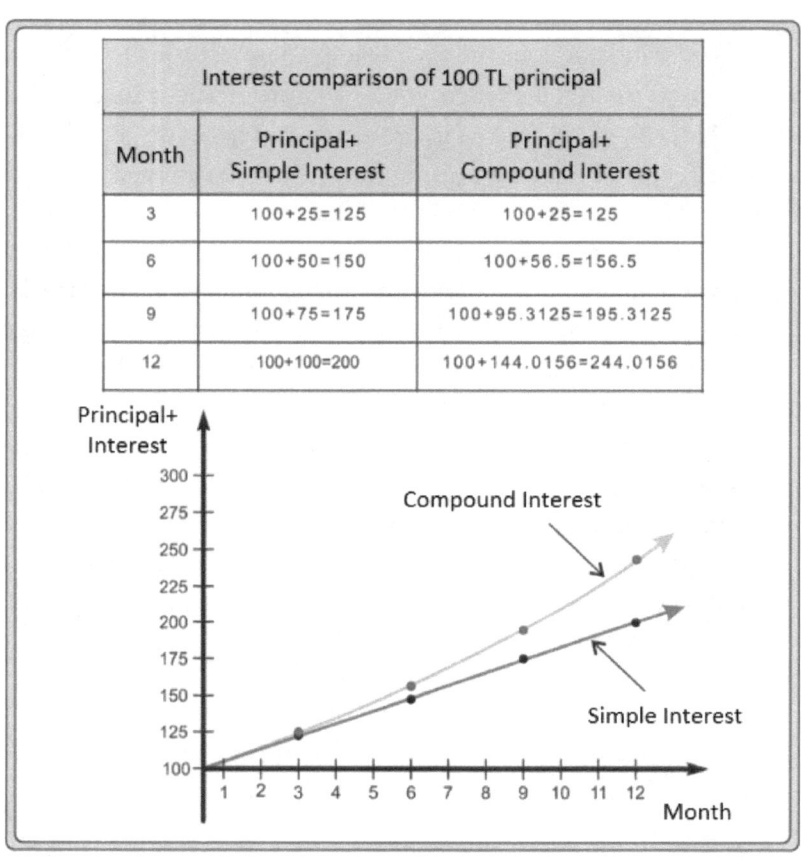

Interest comparison of 100 TL principal		
Month	Principal+ Simple Interest	Principal+ Compound Interest
3	100+25=125	100+25=125
6	100+50=150	100+56.5=156.5
9	100+75=175	100+95.3125=195.3125
12	100+100=200	100+144.0156=244.0156

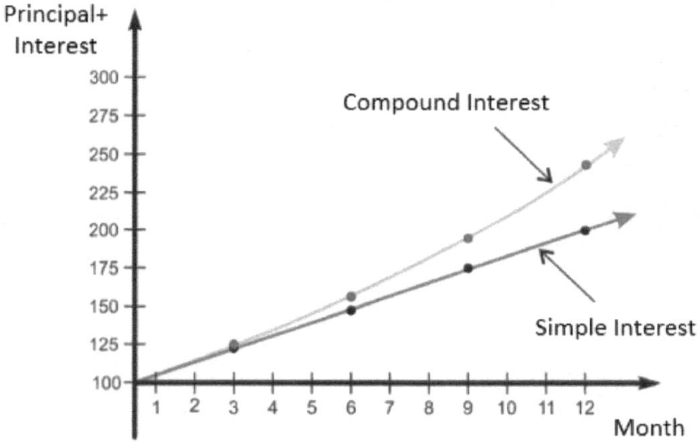

Figure 6.1: Compound interest refers to multiplication together.

As you can see, with simple interest, your 100 liras become 200 liras at the end of the year, which is twice as much, while with compound interest over 4 periods, it becomes 244 liras, which is about 2.44 times as much. Now I would like you to think about what happens if we increase the number of periods to 50, 100 or infinity, in short, what happens to the money we get with instant interest. I have calculated below what happens at the end of 50 periods.

$$(1+1*1/50)(1+1*1/50)(1+1*1/50)(1+1*1/50)(1+1*1/50)$$
$$(1+1*1/50)(1+1*1/50)(1+1*1/50)(1+1*1/50)(1+1*1/50)$$
$$(1+1*1/50)(1+1*1/50)(1+1*1/50)(1+1*1/50)(1+1*1/50)$$
$$(1+1*1/50)(1+1*1/50)(1+1*1/50)(1+1*1/50)(1+1*1/50)$$
$$(1+1*1/50)(1+1*1/50)(1+1*1/50)(1+1*1/50)(1+1*1/50)$$
$$(1+1*1/50)(1+1*1/50)(1+1*1/50)(1+1*1/50)(1+1*1/50)$$
$$(1+1*1/50)(1+1*1/50)(1+1*1/50)(1+1*1/50)(1+1*1/50)$$
$$(1+1*1/50)(1+1*1/50)(1+1*1/50)(1+1*1/50)(1+1*1/50)$$
$$(1+1*1/50)(1+1*1/50)(1+1*1/50)(1+1*1/50)(1+1*1/50)$$
$$(1+1*1/50)(1+1*1/50)(1+1*1/50)(1+1*1/50)(1+1*1/50)$$

$$(1 + 1/50)^{50} = 2{,}691\ldots.$$

I said that simple interest is concerned with the initial state of money in the bank, while compound interest is concerned with the final state. In compound interest, the money at the end of each moment is now the principal. This idea, which gives rise to the number e, tells you that all past accumulation has an effect on the next moment.

Let me say here what I will say at the end: The number e is the name of "Replication Together", which we observe closely in many phenomena in nature. This is why it occupies so much space in scientific life. Since compound interest is the best way to explain the phenomenon of multiplication together, we learn to solve interest problems.

To make the number e a little more visible, if we symbolize the number of periods not as 4 but as n, an expression that goes to infinity;

$$S = s_n * s_{n-1} * s_{n-2} \cdots\cdots\cdots\cdots\cdots s_3 * s_2 * s_1$$

We obtain the expression.

Within this logic $(1 + 1/4)^4$ expression $\left(1+\frac{1}{n}\right)^n$ with the following.

You can easily calculate that n converges to 2,718281828... after two-digit numbers. We call this irrational number "e".

Notice here that in order to reach the number e, we take the interest rate and the coefficient of time elapsed in interest equal to 1. When we were explaining trigonometry, we were using the unit circle to reach the sine and cosine functions. Just like that, this approach, that is, taking 1 as a reference, offers a very important awareness in order to simplify and make operations understandable. With this logic, we can treat the number e as the foundation of a building and easily place other interest rates and durations on this foundation.

In fact, if monthly, weekly or daily compound interest is applied, you can easily see that there is a trend towards the irrational number 2.718281828...... But remember that the real number e can only be reached with "moment"!

In fact, keep in mind that you will never get this irrational number. You cannot perform operations on irrational numbers without rationalizing them. Therefore, I would like to remind you again that operations are performed with the rational numbers to which these numbers converge!

Also $1/n$ instead of $2/n$ if it were $\left(1+\frac{2}{n}\right)^n$ in the formula $n = 2m$ if you write;

$\left(1+\frac{2}{2m}\right)^{2m}$ When you simplify the 2's in the fractional expression in the equation

$((1+1/m)^m)^2 = e^2$ you get the result.

In short, with a simple transformation of the share e'nin you can put it on the base.

In order to establish a functional relationship, the share should be a variable expression. x if you take it, $n = xm$ with transformation;

$\left(1+\frac{x}{xm}\right)^{xm}$ you get.

From here $(1+1/m)^m = e$ because it is,

$((1+1/m)^m)^x = e^x$ you can easily reach equality.

The argument x can be the dependent variable of another expression. If you write x as a function of time, t, and use a coefficient like r to get other expressions, you can write the equation x=rt. And so on;

$((1 + 1/m))^{mrt} = e^{rt}$

If you put an expression with a coefficient at the beginning;

You can obtain $A((1 + 1/m))^{mrt} = Ae^{rt}$

In problems involving the number e, most of the time you do nothing more than finding the coefficients A, r, t. These coefficients expand the number e and pave the way for its use everywhere. Because not everything increases 100% of the time. With these variables, you can represent all functional relationships that increase or decrease together as a function of e.

When describing the number e, we especially use the term moment. Because the expression moment is an expression that gives the number e its soul and it contains a continuity. So, there is no interruption. There is no interruption while events are happening. Now you know that if you take the annual interest daily, hourly or secondly, that is, if you increase the number of periods, the result can be calculated with this coefficient.

But it is important to emphasize here that it is not easy to divide everything into infinity and get out of it, you have to use concepts that come from rationality, which we call rounding and limits. There is a continuum in life, and if you were to videotape it, you would only need to move 16 pictures per second to see it. For a better picture you can increase the number of pictures to 24 or 32. In short, just as you don't need to move infinite pictures per second for video, it is perfectly reasonable to rationalize the number e.

Just as there is multiplication, there is also the opposite, that is, there is also a decrease. Therefore, when perceiving the functional relationship that emerges with the number e, it is necessary to perceive it correctly. While making sense of systems that multiply or decline together, we can make the co-multiplication or decline, which is a part of our daily life called population growth, a function of the number e.

The basic logic here is to find out what the coefficients are in the functional expression we call Ae^{rt}. You can also think of this equation as the function Asin(wx-y). Of course, in such equations, you can add depth to your main functional expressions by assigning a functional relationship to each term such as A,r.t,w,x,y, that is, by assigning variables. The Ae^{rt} I am trying to explain here is the name of the foundation, increasing or decreasing the depth of this is related to your approach to the problem, that is, your mathematical model.

As with compound interest, you can see the relationship between the number e and population growth more clearly when you consider that population growth is driven not only by the people who existed at a certain date, but also by their children and people who migrate. The fact that when a mother gives birth, the person she gives birth to can also give birth at the same time is a good example of co-production.

I would like to emphasize that every birth and death, and dynamics such as out-migration and in-migration, have a direct and immediate impact on this, and that population growth is therefore related to the number e. When I say population growth, do not perceive it only as population growth in cities. You can consider the increase of any size, including plants, animals, cells, workload, labor force or money, within the logic of multiplying together in this way.

Now let's take a look at how the number e and the number π meet in the same movie. Try to understand the subject by remembering that the number π is the ratio of the circumference of a circle to its diameter, and that the number e is related to a ratio of change.

With the introduction of the number π, relating the sine and cosine functions with the number e has facilitated the solution of many physical problems. Of course, polynomials have made a great contribution in proving that they meet somewhere. e^{ix} with $\sin(x)$ and $\cos(x)$ I can say that the fact that when you open the functions to the polynomial, you get expressions that are similar to each other is the basis for establishing this relationship. But when you look at the philosophy of the work, I would like to remind you that the number e and the sine are both based on curvilinear relationships, so it is quite normal that they can transform into each other.

If you want e^{ix} with $\sin(x)$ and $\cos(x)$ Let us show you how the functions are similar to each other through polynomials. First of all

$$e = \left(1 + \frac{1}{n}\right)^n$$

A little bit about equality;

$$e = \left(1 + \frac{1}{n}\right)\left(1 + \frac{1}{n}\right)\left(1 + \frac{1}{n}\right)\left(1 + \frac{1}{n}\right) \dots\dots\dots\dots\dots\dots \left(1 + \frac{1}{n}\right)$$

You may remember from exponents that the above is in the form of a product. By multiplying the two terms in the equation in order, you can deduce what this expression will become.

I wanted to understand this sequential multiplication, $(a + b)^3$ I started by opening his testimony:

$(a + b)(a + b)(a + b) = a^3 + 3a^2b + 3ab^2 + b^3$ as well as this $a\ yerine\ 1,\ b\ yerine\ 1/n$ I would like to say that I see the background of this expansion more clearly when you write and realize that the coefficient in front of each term is arranged according to a rule.

Now e Let's add variables to this expression to obtain functional relationships over the number. If you like e^{ix} Let us continue reading through his statement;

$e^{ix} = \left(1 + \frac{ix}{ixm}\right)^{ixm}$ we can write.

$((1 + 1/m)^m)^{ix} = e^{ix}$ If you polynomialize the equation and rearrange it a bit, you get the following expressions.

$$e^{ix} = [\,1 - \frac{x^2}{2!} + \frac{x^4}{4!} - \frac{x^6}{6!} + \cdots \dots\dots + (-1)^n \frac{x^{2n}}{(2n)!}\,]$$
$$+i\,[\,\left(\frac{x}{1!} - \frac{x^3}{3!} - \frac{x^5}{5!} - \frac{x^7}{7!} \dots\dots (-1)^n \frac{x^{2n+1}}{(2n+1)!}\right)\,]$$

Similarly, $\sin(x)$ and $\cos(x)$ when you open the expressions to the polynomial;

$$\sin(x)=\frac{1x}{1!}-\frac{x^3}{3!}+\frac{x^5}{5!}-\frac{x^7}{7!}+\quad \dots \qquad +(-1)^n\frac{x^{2n+1}}{(2n+1)!}$$

$$\cos(x)=\frac{1}{1!}-\frac{x^2}{2!}+\frac{x^4}{4!}-\frac{x^6}{6!}+\quad \dots \qquad +(-1)^n\frac{x^{2n}}{(2n)!}$$

You get their equality.

$\sin(x)$ fonksiyonunu i multiplied by

$$\text{isin}(x)=i\left[\frac{1x}{1!}-\frac{x^3}{3!}+\frac{x^5}{5!}-\frac{x^7}{7!}+\quad \dots \qquad +(-1)^n\frac{x^{2n+1}}{(2n+1)!}\right]$$

we get.

The equality is actually the result of a multiplication operation, and when you organize these operations a bit, you can see the picture a bit more clearly with the factorial and combinatorial expressions above.

You felt a bit uncomfortable when a complex number entered into the equation, didn't you? Complex number is actually a number system that has entered our lives because it expands the number system and therefore the solution set of the problem. As you know, complex numbers include real numbers and every real number is actually a complex number without a virtual part. Since it has the ability to extend the number system directly to the surface, complex numbers are used to find solutions to more problems.

Now e^{ix} Let's clear up some of the confusion caused by the complex number in it. The most basic complex number notation $a+ib$ When you expand the expression into a power series, you again get a binary system consisting of real and imaginary. $(a+ib)^2$With the expansion of $a^2+i2ab+b^2$you get the expression. You can also use this $(a^2+b^2)+i2ab$ You can write real and imaginary numbers side by side. iSince by definition the odd powers of are imaginary numbers and the even powers are real numbers, these expressions naturally arise by themselves. Just like this e^{ix} It is quite plausible that real numbers on one side and imaginary numbers on the other.

This book is not a textbook, so I have not explained in detail how they are derived here, but you can see that they are derived in a similar way if you do a little math. Now look at these expansions; how similar are they? e^{ix} And the relationship between sin(x) and cos(x) becomes even more obvious when you look carefully at the series, right?

$$e^{ix} = cosx + isinx$$

e^{ix} one side of the equation that gives the expression $cosx$ one side $sinx$ function is no longer so difficult. Known as Euler's formula $e^{ix} = cosx + isinx$ You can see that polynomials, i.e. series, come to the rescue in the emergence of equality.

x instead of π If we put $e^{i\pi} = cos\pi + isin\pi$ you get the expression. Here we get $cos\pi = -1\ ve\ isin\pi = 0$ because it is;

$e^{i\pi} + 1 = 0$ you get the expression.

Sine, cosine, e, logarithm, series, sequences, binomial, polynomial, etc., have a purpose. These are concepts created and derived so that you can analyze this life with mathematics. The more you use these concepts, the more clearly you will see what mathematics wants to tell you.

Co-replicating systems are very common in engineering and, of course, in many parts of social life. In everyday life, for example, we can explain the spread of news with the logic of co-replication. Now you tell a news to one person and he tells it to another person. If, in this process, the person who tells the news stops telling the news after telling one person, we can say that there is simple interest here, but if the telling turns into a continuous and joint activity, we can say that there is compound interest here. You can see the spirit of the number e if everyone is a partner in this process along with the people you tell, that is, if everyone works to spread the news.

You can see the traces of the number e behind the tragedies that sometimes occur in the stands and during the stoning of the devil in the Hajj. Since people cannot see the power and magnitude in compound interest as clearly as they can in simple interest, they cannot see the catastrophe and events become unavoidable. While the total weight and pressure of people moving or standing at constant speeds can be perceived more concretely, in a state of panic, it is not perceived that maximum pressure is applied to the weakest area in a very short time due to the effect of acceleration and multiplication together, which is why such disasters occur. You can see the spirit of the number e behind the creation of this pressure. If you catch this logic, you can find many different examples of co-proliferation and derive their mathematical models.

Let's continue explaining this subject with a more concrete example. Let's take a look at how the number e is related to the process in which the apprentices assigned to masters in a factory learn the job and help their masters, who in turn train new apprentices and bring them into the workforce. Keep in mind that the productivity of the masters, the timing of apprentices starting work and the rate at which people leave the business all have the power to change the coefficient on the e number.

Here you will get a function like $f(t) = Ae^{rt}$ and the coefficients A, r, t tells you what your function is. When you find them, you have already solved the problem. Let's continue explaining this with a simple example:

Employees	QTY	Monthly parachute production in 2018											
		1	2	3	4	5	6	7	8	9	10	11	12
Master	50	125	95	85	96	80	90	75	125	110	114	125	134
Apprentice-1	20				86	85	68	95	78	95	86	45	75
Apprentice-2	16								85	58	76	49	78
Apprentice-3	14										89	64	25
Total	100	125	95	85	182	165	158	170	288	263	365	283	312
Parachutes produced in 20 years	40.000	40.125	40.220	40.305	40.487	40.652	40.810	40.980	41.268	41.531	41.896	41.179	41.491

Figure 6.2: A simple data table to illustrate the use of the number e.

A parachute factory employs 50 staff. These employees have made 40,000 parachutes in the last 20 years. In 2018, the number of parachutes made by 50 masters on a monthly basis is given on the next page. In 2018, with the start of new workers in the factory, these personnel were also involved in the production process after a certain period of time, and the parachutes made by them are also given on the next page. In 2018, the total number of parachutes made in addition to 40,000 parachutes at the end of each month is given in the bottom row. Now, when you look at these figures, let's solve the problem by explaining how many parachutes were made in which month and how many parachutes were made and how the estimated number of parachutes for 2019 is related to the number e.

First of all, you need to analyze the table well. It is worth emphasizing once again here. Almost all of the formulas are derived from experimental results. Look at the table above in this light. So first of all, you need to appreciate the database you have.

Now let's estimate the production in month 12 or month 24 based on the increase between month 3 and month 10:

First of all, we need to find the most realistic function that passes through the resulting points by transferring the data we have to a coordinate system. The best method for this is the polynomial approach. While the master continues production, the apprentice he trained joins the production after a certain period of time and the new apprentices he trained join the joint production. Here we can talk about the principle of co-proliferation and therefore a situation related to the number e. Now it goes without saying that you need polynomials that can be written using the rate of increase and the concept of time. Here is the function $f(t) = Ae^{rt}$ where they work together and the most appropriate expression in the spirit of the increment is the function $f(t) = Ae^{rt}$.

$f(t) = A\left(1 + \frac{rt}{n}\right)^n$ when you open the expression

$f(t) = Ae^{rt}$ function.

When you transfer the table above to the coordinate axis, this function will plot the closest curve passing through the intersection points of the production data corresponding to each month. Now let's try to find the coefficients A, r, t from the sample data:

$f(3) = Ae^{3r} = 40305, \qquad f(10) = Ae^{10r} = 41.896$

That is, you can deduct from the table and when you proportion them;

$\frac{Ae^{10r}}{Ae^{3r}} = \frac{41.896}{40.305}$ with $e^{7r} = 1{,}04$ you'll find it.

If you take the natural logarithm of both sides;

$7r = 0{,}0387$ you'll find it. From here $r = 0{,}00553$. you will find

Now $f(3) = Ae^{3r} = 40.305$ equality;

$A = 39{,}641$ you'll find it.

Actually, A=40,000 should have come out. Since this is an estimate, it is normal to get such results. We are also trying to find a rough solution by making some simple assumptions without adding a deeper meaning to this functional relationship. Remember, you cannot solve any problem without rounding. Now;

$$f(12) = Ae^{12r}$$

When you substitute A and r values in the equation,

$f(12) = 39{,}641e^{12*0{,}00553}$ you'll get it.

And from there $f(12) = 42.361$ you'll find it.

Look at the real value;

$$f(12) = 42.491$$

we're very close to the real result, aren't we?

$f(24) = 45.268$ you'll find it.

By the end of 2019, production will be about this amount.

You look at past data and make a prediction about the future.

This is an estimate, remember! Our result is very close to reality, but of course we need more data to make the most accurate prediction.

As a result, the number e is a very important number in understanding and analyzing many situations in this life. It is possible to encounter the number e in all kinds of problems such as the interest of interest, or the inclusion of children in this process over time in population growth, or the spread of news.

Imagine a pressurized environment. When you add additional force to a system under the influence of pressure, the forces start to increase the pressure of the environment together. When you imagine an environment where this happens continuously, you can reach the number e from this.

The number e is used wherever we speak of multiplication together. Of course, it is also used for decreasing together. Of course, you can easily fill in the table about parachute production for any problem. Then you can see the number e in many parts of this life.

I hope you have a better understanding of why I emphasize the number e so much. π Like the number e, the number e, which is a symbol of multiplication together, should also have a day. In my opinion, the most appropriate date for e-day is September 28th. Because this day is the 271st day of the year and the 272nd day every four years. You can associate this day with the first digits of the number e, which continues with 2.718.... I think you should think about it!

Limits and Series

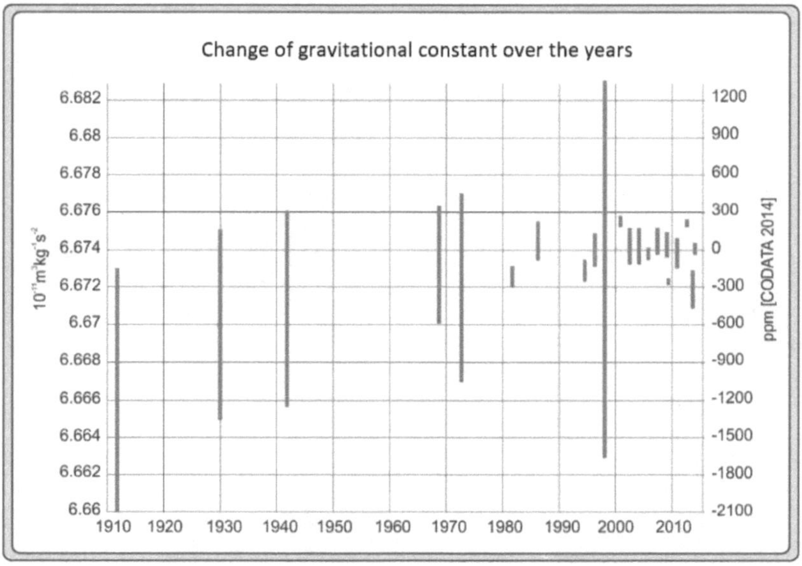

Limits and Series: Turning an Irrational Life into a Rational One Bring

Before explaining the limit, I would like to talk about the concept of rounding. When I first learned about rounding in math classes, I remember saying, "Why are we rounding, I want to find the exact result!" Is it possible to make a scientific discovery or a technological product with incomplete or half calculations, right?

"If we could calculate everything exactly and precisely, we could build technological products such as airplanes, computers and satellites," I might have thought at the time. Nowadays, I say, "Without rounding, there would be no airplanes, computers or satellites!"

In fact, we do round everywhere in life. Now take a tape measure and measure the width and length of your room. If forty people measured the room, everyone would give a different result, wouldn't they? Many variables cause this, from the precision of the measuring instrument to the skill of the measurer and the location of the points you use as reference for measurement. Rounding gets rid of these uncertainties. Also, if you don't stop measuring at some point, you will have to measure the width and length of the room until the morning.

If you want to have meaningful information, you need to stop measuring somewhere, write down the numbers on paper and start solving the problem. No matter what you write on the paper, the numbers will still be rounded. If you wrote the width as 3.165 m, maybe the width was 3.165323..., but you couldn't measure that precisely and stopped at 3.165. So, rounding gives you concrete expressions that lead to a tangible result.

There is an irrationality in measurement. In reality, you can never arrive at a clear number as a result of measurement. Since we cut out these infinite numbers somewhere, we can make them talk to each other, add, multiply, divide and subtract them. In other words, by rationalizing an irrational life, we make it meaningful and solve problems.

When you build a mathematical model of a real problem, you are actually doing a rounding process. The more realistic your model is, the more realistic the results will be. This conformity depends on your rounding precision. For example, you are building a bridge and you need to calculate the loads on its two legs. As you transfer the bridge to paper, you need to round many values from the shape and dimensions of your geometry to the quality of your concrete, from the iron you use to the foundation of the bridge piers

Another name for rounding is optimization. You can even call the concept of optimization a college graduate version of rounding. We always optimize when solving problems. When solving a problem, you first need to create a mathematical model of it. This model should have a feature that describes the reality in the closest way and does not detach the problem from its soul with unnecessary information. Optimization is another name for optimizing a task at hand. Any kind of assumption made while optimizing is actually a rounding process. In short, you can neither optimize nor solve a problem without rounding.

The most important mathematical operation associated with the concept of rounding is the limit. At the beginning of the book, we said that analytical thinking is based on induction and deduction. The basic philosophy of analytical thinking is to break a problem into small pieces and then add them up to get the big picture. I will explain it in more detail in the future, but let's give a cursory overview here; while the integral allows you to reach the big picture by combining these infinitesimal pieces, the derivative allows you to find out what each of these infinitesimal pieces means. Here, the limit provides the unity of meaning between these parts. You get rid of the uncertainty of infinity thanks to the limit. The limit determines the boundary of infinity.

Infinity actually means uncertainty. Uncertainty also means an insolvability. We learn mathematics to make uncertainties certain. The product of zero and infinity ($0 \cdot \infty$) yields one of the most valuable uncertainties. On the one hand, an infinitely small part, on the other, an infinitely large part. From the product of these two, meaningful expressions are derived. Yes! Where do we see these two concepts that come to mind when we think of "infinity and zero"? Take a look at the integral. The widths of rectangles go towards zero, and these rectangles are infinitely many. In fact, deriving a meaningful result from an uncertainty is a great solution that mathematics offers us.

In the same way, we can talk about a similar uncertainty in the derivative. The derivative is described as the ratio of two sides. When your ratio two infinitesimal lengths, you open a door to the derivative. There is uncertainty here too, isn't there? We can call each of the two infinitesimal quantities zero. When you divide them by each other, that is $0/0$ I want you to see that a meaningful concept like derivative emerges as a result of the expression.

Normally, you cannot directly derive from the $0/0$ uncertainty. This is where the limit comes in and allows you to draw meaningful conclusions from such uncertainties. Once you know this, you will have a better understanding of what the limit means and why it is so important for mathematics.

With limits, we overcome the uncertainties that we encounter directly with the most important topics of mathematics such as derivatives and integrals, and transform them into meaningful expressions. If there were no limits, we would not have been introduced to two fundamental mathematical teachings such as derivatives and integrals, which are the most widely used in the development of scientific thought.

If you can make sense of these ambiguities, of course you can make sense of other ambiguities. For example, when you solve ambiguities like ∞/∞ or $0/\infty$, you solve other problems. What I am trying to explain here is the fact that we provide the unity of meaning between infinitely small and infinitely large parts through limits. In short, we make an irrational world rational through limits.

I remember in high school, when the concept of limit was explained, we were told that if you travel half as far each day as the previous day, you will never reach your goal. It was also mentioned that this problem is solved thanks to the limit. For example, you have a 2 km road ahead of you and you traveled 1 km on the first day; if you continue in this way every day, traveling 1/2 km on the second day and 1/4 km on the third day, you can see that you will never finish the 2 km long road. Without the concept of limit and infinity, we would have great difficulty in perceiving these things.

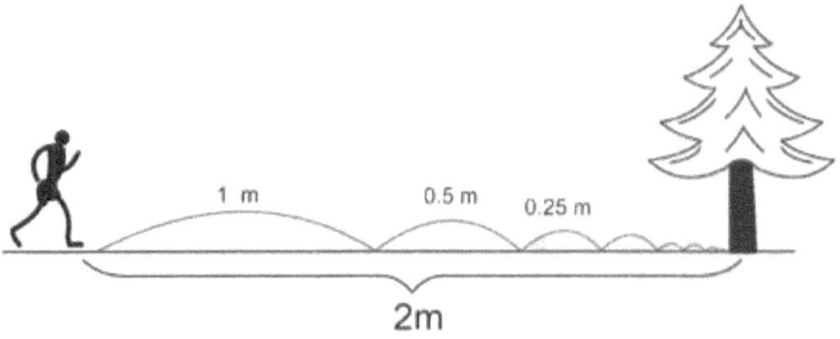

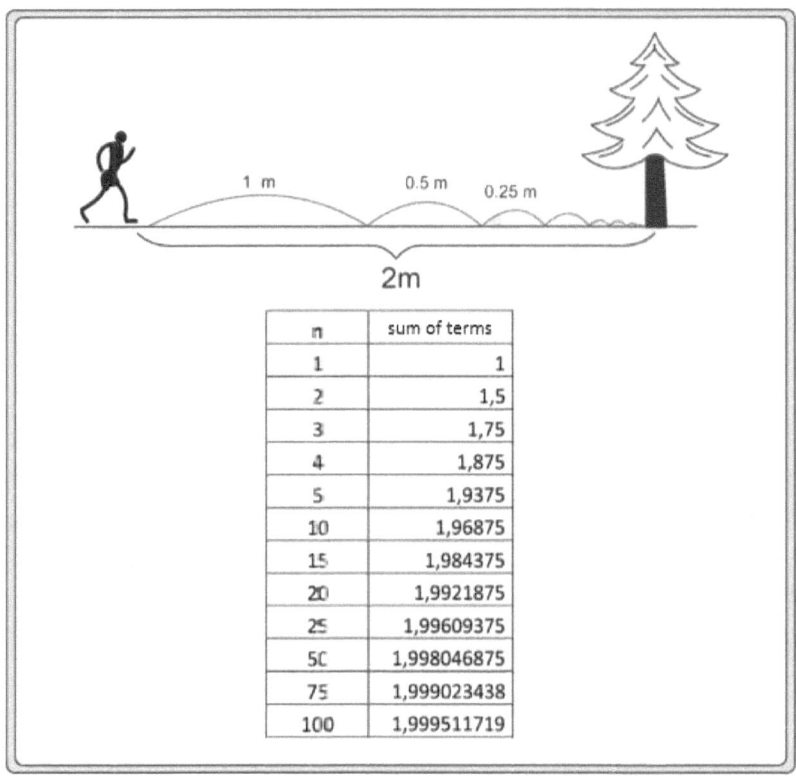

n	sum of terms
1	1
2	1,5
3	1,75
4	1,875
5	1,9375
10	1,96875
15	1,984375
20	1,9921875
25	1,99609375
50	1,998046875
75	1,999023438
100	1,999511719

Figure 7.1: Thanks to the limit, we use tangible quantities.

In fact, this road story is a good example to grasp the concept of limit. However, since we cannot see this in other areas of our lives, I can say that the concept of limit is not very imaginary and has no place in our real world. When you bring the subject to derivatives and integrals and their relationship with real problems in life, the concept of limit will tell you its true meaning, why you need it and why it should be learned.

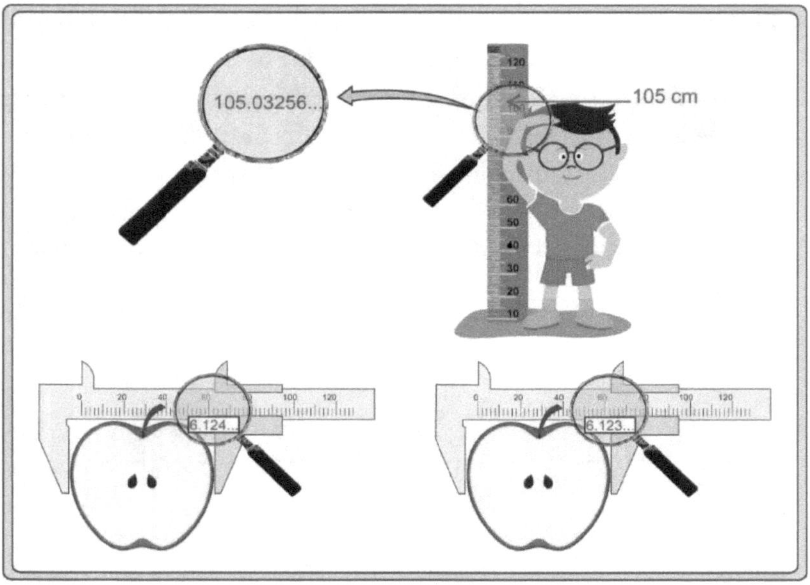

Figure 7.2: Measurement is where you rationalize irrationality.

I have already said that the basis of scientific thinking is measurement, and that when making measurements, you have to constantly round off, in other words, you have to take limits. The results of measurements fill the set of data, but it is the limits that turn them into meaningful expressions. You can never measure your height or weight exactly. You can never divide an apple into 2 or 3 equal parts. You can never tell exactly what time it is. You always take the limit of the value you measure and tell the result. In short, if there is a measurement, there is a limit. You cannot make scientific calculations without limits.

Now let's start analyzing the path problem I mentioned earlier. Sum the table on the previous page ($\sum$) symbol, we get the following expression:

$$\sum_{i=0}^{n} \frac{1}{2^n} = 1 + 1/2 + 1/2^2 + 1/2^3 + 1/2^4 + 1/2^5 \ldots\ldots + 1/2^n$$

You cannot spend a lifetime adding up the numbers above, one term at a time, forever. Now let us continue to understand what I mean by adding their first terms one by one:

When you take the first 5 terms, the result will be 1 + 1/2 + 1/4 + 1/8 + 1/16 = 1.9375, while when you increase the number of terms collected to 10 terms, you reach 1.999. Now look carefully at these two sums;

$\frac{1}{2^0} + \frac{1}{2^1} + \frac{1}{2^2} + \frac{1}{2^3} + \frac{1}{2^4} + \frac{1}{2^5} \ldots + \frac{1}{2^n}$ (1) you get the expression.

You can see that it adds up to 2, right? Thanks to the limit, you have a tangible result, you can use it wherever you want.

Now let's divide the number 1 by 3: 1/3 = 0.333333333... and you will see an expression that goes to infinity, right?

This statement;

$\frac{1}{3} = \frac{3}{10} + \frac{3}{10^2} + \frac{3}{10^3} + \ldots + \frac{3}{10^n}$ (2) we can show it as.

$$\frac{1}{3} = 0{,}3 \quad 3 \qquad 3 \qquad 3 \qquad 3 \ldots$$

$$\frac{1}{3} = \frac{3}{10} + \frac{3}{10^2} + \frac{3}{10^3} + \frac{3}{10^4} + \frac{3}{10^5} \ldots$$

Then the opposite;

$\frac{3}{10} + \frac{3}{10^2} + \frac{3}{10^3} + \frac{3}{10^4} \ldots + \frac{3}{10^n} = \frac{1}{3}$ you have already proved it from the beginning.

If we visualize this a little more;

$$\sum_{i=1}^{n} \frac{3}{10^n} = \frac{1}{3}$$

I think it goes without saying. Whether you look at the first or the second statement, I want you to see that they both say the same thing on the way to infinity. Of course, you can find many more examples like this in the number set.

Note that when you divide 1 by 7 you actually get a series, and similarly 19/7 and 22/7 I explain that division operations like this can be opened in the same way as before. As you know, these two division operations "e" and "π "numbers. But as a result of all this "e" and "π" numbers are obtained, they are just an approximation.

I use the term convergence specifically here. The limit is actually a convergence. As you know, if you want to get meaningful expressions from a data set that you can transfer to a coordinate system, you first need to understand the curve in the coordinate system. I have mentioned this before: Life does not always present us with linear relationships; in fact, almost everything in this life is connected by curvilinear relationships. We make sense of them by breaking them down into line segments. Isn't Linear Algebra already the art of making lines out of curves?

The complexity of life comes from curvature, whoever can extract the most truth in this life will succeed. Curves are so important that curves are at the heart of engineering problems and mathematics has become richer to analyze them. One of the best ways to understand these curves is through series.

You may have noticed the complexity of logarithmic and trigonometric functions. Because of this complex structure, we find it difficult to perceive them. When solving a problem related to them, it is not so easy to get out of it. In order to obtain meaningful expressions from them, we use series, which are expressions that expand to infinity. Without series, it is almost impossible to analyze them.

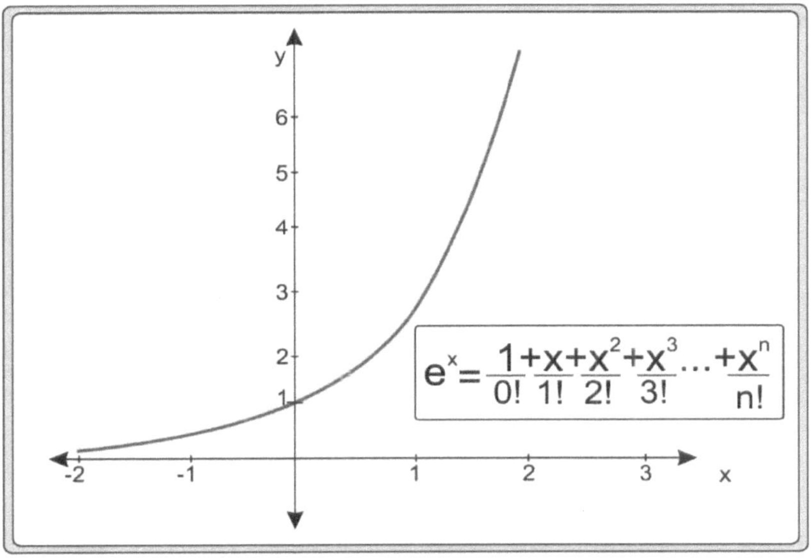

Figure 7.3: The series allows you to clearly see each of the line segments that make up the curve.

Functions have a special place in mathematics. As I have mentioned before, it is thanks to functions that we can transfer the facts of life onto paper. If you transfer the functions to the coordinate system, draw their graphs and analyze them, you will solve a large part of the problem you are dealing with.

Approaching the graph of a function in a coordinate system, in other words, its curve, with expressions whose coefficients are certain, makes it very easy to analyze the curve. We call expressions with certain coefficients *polynomials*. Thanks to these polynomials, once we have made the curve expressible in meaningful numbers, we can play with it like a ball. Each term in the polynomial describing the curve has a clear equivalent and it is not very difficult to take their derivatives and integrals. These polynomials open the door to series.

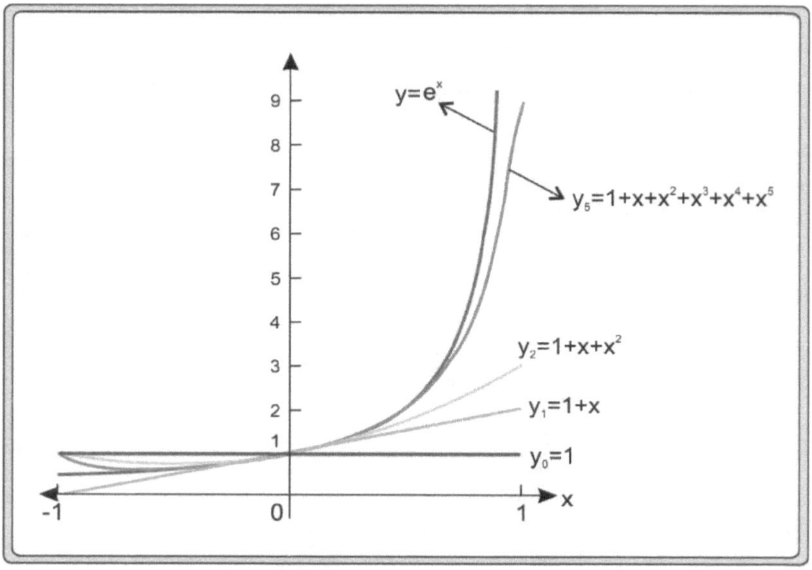

Figure 7.4: The series guides you to see the main picture.

When I explain derivatives and integrals, I will be explaining how we take the derivatives and integrals of expressions like x^2 and x^3 . Taking the derivatives and integrals of these expressions is much easier than taking the derivatives and integrals of a logarithmic or trigonometric function. This is why series, which are also described as unfolding polynomials, provide us with a great deal of ease of operation.

Thanks to rounding and limits, uncertainties in equations are eliminated and meaningful results are obtained. The above examples tell us that the problems in the physical world are concretized and transformed into meaningful numbers thanks to the concept of limit. Mathematics does not like contradictions within itself. Look at limits and series with this perspective. While learning mathematical concepts, keep in mind that you need to approach the subject with integrity.

If you want derivatives and integrals, the most commonly used functions in mathematical analysis, to tell you something, you need to learn the concepts that make them meaningful. There is a basic approach to the integral and derivative functions: Break the problem into very small pieces and then put them together in a neat way. If you want very small pieces to turn into meaningful expressions, series will help you a lot.

For curves to be analyzed, there must be no breaks in the shape. This is like an open circuit in electricity. You can have as many cables as you want, but if there is a break or a lack of contact in the cable, you cannot transmit the electricity to where you want. We call this *continuity* in mathematics. A series that leads to a curve must be continuous. If a function is not continuous, you cannot perform an operation, you cannot start with it, it will not tell you anything clear, it will let you down. Series are not a balm for the wound.

If we explain continuity through a simple example;

$$1 + 1/2 + 1/4 + 1/8 + 1/16 \ldots\ldots. +1/2^n = 2$$

subtract a few initial terms from the sum.

or;

$$1/3 = 0{,}333333 \ldots.$$

When you subtract the third and fifth decimal terms from the expression, you can see that the sum corresponds to a completely different number. You can neither reach 2 nor 1/3 with these sums. Continuity ensures that there is no such discontinuity.

To better understand continuity, it is first necessary to understand what a discontinuity is. Discontinuity is like scratches on a CD (Compact Disk). If there are scratches on a CD, that CD will not work. In the same way, if there is no continuity in your functions, that function will not take you in the right direction.

Do not look for an eternity in continuity by looking at the n-words above. Of course, you can also use continuity at certain intervals. For example, if you say "I need a 5-minute image", you can define a continuum that will give an image of that duration. In other words, the absence of scratches in that area of the CD will be enough to watch that image.

Continuity is a concept created for us to understand life. We live a life as if we are having a photograph taken every moment. When we combine these moments, that is, when we make them continuous, we get meaningful images from these pictures. In fact, an image is nothing but the movement of pictures at a speed that the eye cannot understand. Take a look at the first movies that were made, and you will see that images were obtained when the photographs were played back at 16-24 frames per second!

Perhaps we are playing our part on a stage where 1000 pictures are taken every second. When we combine these pictures, we call the image we get life. In fact, we may be living a life here, flowing in an intermittent time. With quantum physics, the idea that it is possible to travel through time was proposed. That's why movies about the future and flashbacks were made, isn't it? It's like flipping through the pages of a book. We can put the pieces together with time-dependent continuity functions. If it is possible to move between these pictures, would there be an apocalypse if we put this into functions and look for ways to move through time?

While analyzing curves, mathematicians have found many different series according to their needs. I will not tell you about all of them one by one. Here I would like to talk about the most commonly used power series in engineering problems. In fact, the power series can be considered the ancestor of other series. The most basic representation is given below. The first equality is defined as the most basic representation of the power series if we are going to open the series around a fixed point, and the second equality if we are not.

$$f(x) = \sum_{n=1}^{\infty} a_n (x-c)^n \; veya \; f(x) = \sum_{n=1}^{\infty} a_n x^n$$

x variable, a_n are fixed numbers and c is any fixed number. The most valuable observation to note here is that the power series expansion is a sum expressed in polynomials where the natural numbers meet the four operations.

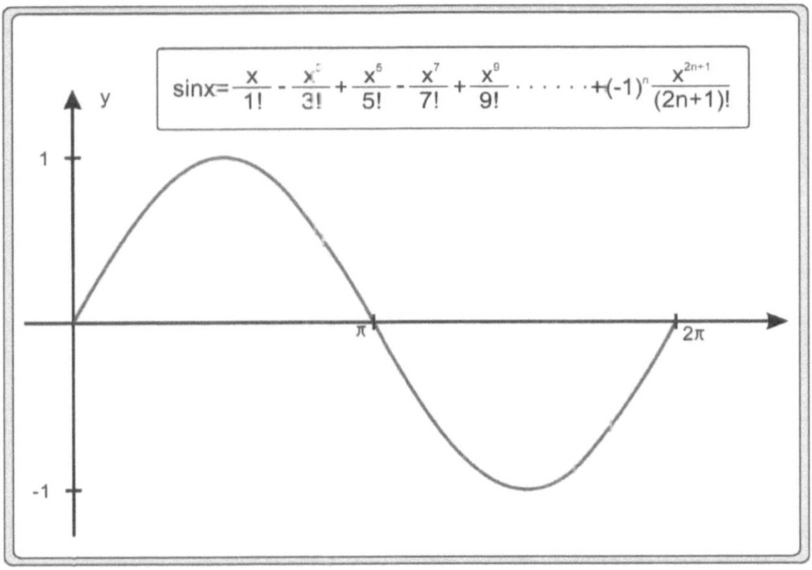

$\sin x = \frac{x}{1!} - \frac{x^3}{3!} + \frac{x^5}{5!} - \frac{x^7}{7!} + \frac{x^9}{9!} \cdots\cdots + (-1)^n \frac{x^{2n+1}}{(2n+1)!}$
y
1
-1
π
2π

Figure 7.5: Series expansion of the sine(x)f function.

In summary, your series must be convergent and continuous to reach the curve. Because if the series is divergent, it is as if it is moving away from the subject; if it is discrete, that is, if it is not continuous, functional integrity is not ensured and it brings completely different values as a result.

Power series are expressions where you can see the polynomials very clearly and they are written as powers of a variable. The main goal of power series is to find the most accurate polynomial sequence that will give the curve of the function in the coordinate system. The great thing about opening a function to a power series, i.e. polynomials, is that you can easily add, subtract, divide and multiply with them.

Decomposing a function into power series or, in other words, polynomials has enabled many functional relationships to emerge spontaneously and thus to see the solution of problems.

The power and leading coefficient of each term of the polynomial is integers. This is to avoid ambiguity and confusion. In short, each term of the polynomial is clear, not irrational. Therefore, you will not get confused.

Each term of the polynomial corresponds to a point on the coordinate axis and you can connect them with lines. If you like, graph a function expanded to a polynomial on the coordinate axis. You will see that a graph that looks like a curve from a distance actually corresponds to a line segment when you look closely.

As you know, we said that the pictures of functions in the coordinate system often appear as curves. In order to make these curves comprehensible, I explained that series are very useful. These curve-shaped pictures, which appear complex in the coordinate system, get rid of complexity thanks to series. Thanks to series, seemingly complex functions can be added, subtracted, multiplied and divided.

If you are asking, "Where did these curves come from?" I would like to remind you again that they mostly come from experimental results. When you do an experiment and transfer the results to the x and y axis, that's when you enter the world of curves. To build meaningful functional relationships from these curves, you need to be able to write the closest functions that describe the curves. In order to obtain these curves, power series of polynomials, i.e. expressions with clear coefficients, will provide you with great convenience.

Getting the curve is actually not that easy. Now I want to give a small example of how a function converges to a curve by defining it with a power series.

First, take two functions, one consisting only of a polynomial of higher degree and the other being the sum of a polynomial of higher degree and two polynomials of lower degree.

Now take a good look at the curves they form. Look carefully at the second function and you will see that the low-order polynomial is trying to bend the neck of the high-order polynomial, the high-order polynomial gives the main picture and the low-order polynomial gives the small corrections. In fact, the values corresponding to the y-axis tell you the same thing. You can easily see here how a high degree polynomial interacts with a low degree polynomial.

Let's explain a little more; let's give x a value from 1 to 5 and observe the change in the y value in both functions. The results here will tell you that the higher degree polynomial produces the main curve, while the lower degree polynomial tries to give a small correction. You can, of course, reproduce them and observe how the degree of the polynomial interacts with the shape of the functions.

Let these be the following functions for example. Draw the graphs, then observe how the shape of the functions y_1 and y_2

changes as x changes; then you will see what I mean.

$$y_1 = x^5$$
$$y_2 = x + x^5$$

I took a course on artificial neural networks at university, and the course was about how machine learning can be. There, I realized that you have a target curve, and to reach it, you draw a curve according to your mind, and then you compare these two curves and go towards the actual curve by changing the coefficients that make up your curve. I'm explaining the same thing here. I'm trying to give you an idea of what the degrees and coefficients of polynomials can mean and how they can bring you closer to the actual curve.

Draw a function with a large polynomial and a small polynomial together and see how the curvature changes. When you define a function by opening it to a series or, in other words, by opening it to a polynomial, you can draw very meaningful conclusions from it. Here you need to be able to intuit which polynomial can shape the curve and how.

Some functional relationships may be very difficult to analyze, but when you see that they form curvilinear relationships, you will see that series produce the closest solution to analyze these curves. This is because the spirit of mathematics, which is divide, conquer, manage, in other words induction-deduction, also manifests itself here. We make sense of big pieces by breaking them into smaller pieces. We are all very familiar with $f = ma$, $e = mc^2$ I would like to remind you again that formulas such as these are a summation, that is, induction, and that they are often broken down and meaningful expressions are sought in their details.

Don't just read series in terms of power series, it is also possible to turn a function into a series by taking its derivatives. Do not think of complex expressions when you think of derivatives. I will explain what the derivative means in the following sections, but let's talk a little more about it here;

$y = x^n$ derivative of a polynomial $y' = nx^{n-1}$ you remember from the lessons.

The above statements tell you that what you call a derivative is also a polynomial or a function in general. If the derivative is also a function, then it is possible to decompose your functions by adding them together. Just as apples and oranges cannot be added together, we would not be able to add them together if the derivative and the integral were not functions. Then it follows that all functional relationships, whether logarithmic or trigonometric, can be taken into the same picture thanks to polynomials as long as you do not break the rules. Therefore, four operations can be done with all of them. As you know, such functional equations are called differential equations.

The subject called differential equations comes up a lot in university education. The common feature of these equations is that the function itself, its derivative and the integral are in the same functional relationship. Such relationships may sound complicated. But the derivative and integral of a function is also a function, as long as you remember that, they don't seem complicated. Therefore, differential equations contain the fact that a function is nothing but addition, subtraction, division and multiplication. Whether it's a trigonometric function or its derivative or integral, whether it's a logarithmic function, the series tells us that they are all the same kind.

Remember, almost all experimental results do not make sense without rounding. When you see that you need to use concepts such as rounding, limits, continuity to get the closest picture to the picture that a function creates in the coordinate system, and that, of course, series, together with concepts such as rounding, limits, continuity, transform functional relationships into the most meaningful expressions, you will better understand the value of these concepts. So, know the value of series very well.

I tried to explain that basic concepts such as series, limit and continuity emerged to rationalize the problems we face in life. The fact that the decimal numbers that go to infinity in the expression 0.33333..., which emerges with 1/3, is a series expansion and that the resulting series expression leads to 1/3 thanks to the limit. I also tried to explain that all functional relationships can be explained by the main idea of dividing series into small parts.

Although we cannot directly relate concepts such as series, limits and continuity to physical and chemical phenomena, they have a lot of mathematical meaning, and without these concepts we would not understand derivatives and integrals. If you look at the reality behind the concepts in this way and learn the need and reasons behind them, you will know where and for what they are used, and you will reach the answers much faster.

Integral

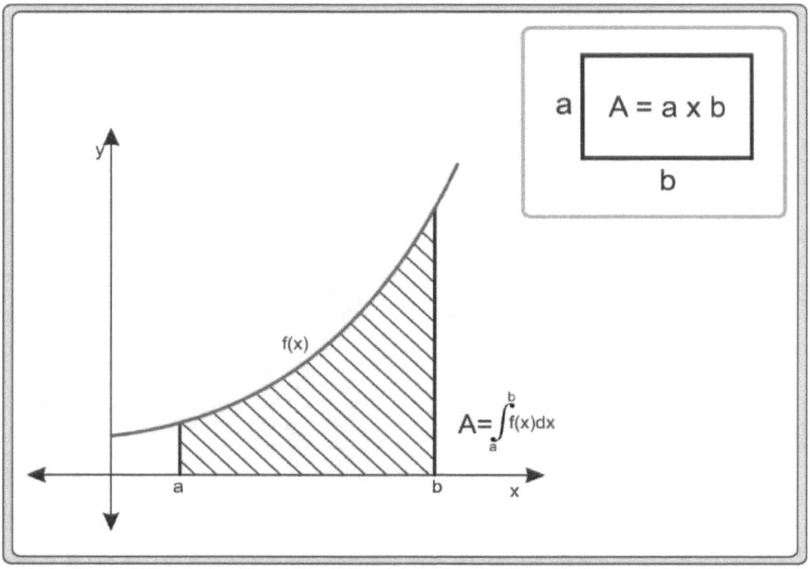

What does it mean to take an integral and what does it do?

The integral is known as one of the most difficult topics in mathematics. Most people who are interested in mathematics look at the integral as the nirvana of mathematics. Like people who love mathematics, I was very excited when I was a senior in high school when I was told about this subject. I don't know if it was because of its name, but this subject was told like an urban legend. I was eager to learn it and when the lessons were over, I thought I had learned it very well.

When I finished high school and went to university, I don't remember almost any course in which the integral was not used. Most of the courses I took involved integral calculus. Since I had learned it very well in high school, I was solving all kinds of problems related to the subject and walking around saying, "I have mastered the subject!"

In high school we were told that integral simply means sum. Although I heard and knew this concept all the time, if you take out the parts I memorized, I can say that I graduated from high school and university without learning where and why the integral is used.

When I was learning the integral, I even remember saying, "Where did this freak symbol come from when we have only one sum symbol?" I could solve integral problems very well, but I did not have the answers to questions like "Where, how and why is the integral used?".

Believe me, if a survey were conducted in our country asking "What is the percentage of students who use integrals after their university education?", the results would be disastrous. If we announced the results, students would be demoralized and would stop listening to the lectures. Unfortunately, just as most of the subjects taught in our education system have no relevance in our lives, neither does the integral.

In schools, we keep stuffing millions of students into classrooms and teaching them mathematical concepts that come out of life without giving them the spirit of mathematical concepts, without telling them what they can do when they learn them, but only by memorizing the rules. Of course, because of this fiction, the integral cannot save itself from being a subject whose rules are memorized.

Why do people who deal with previously solved problems by looking at the answer sheet need to learn the integral! There is no point in teaching us the integral except to give us a general cultural knowledge for analytical thinking. Unfortunately, we are not a country that offers much of an ecosystem for people to work in if they want to learn the root of the matter. Therefore, people who learn these subjects immediately go abroad to use what they have learned.

When you read this, don't immediately say, "Let's remove these subjects from the education system!" Even though we teach only the rules of the integral, knowing the integral has an important contribution to scientific studies in the world. The author of this book first memorized the rules and now he interprets them. If the integral is to be taught, this is why it should be taught, after all, we are a developing country. In the future, more people will start to question why they learn what they memorize, and perhaps our educational structure will change radically.

I think it goes without saying that the integral, like all other operations in mathematics, emerged out of necessity. The number of mathematical concepts increased logarithmically with the beginning of the enlightenment period with the use of analytical thinking, in other words engineering thinking, in solving problems in life.

When you look at the processes by which these concepts are generated and derived, you will see that most of the time, we are looking for ways to break large parts of the problem into small meaningful parts and to reach large parts from them. The integral is a mathematical function created to serve this purpose.

Humans invented the microscope and telescope to see and understand small and large pieces. Just as the telescope and microscope were invented to make sense out of the pictures of small and large pieces, the concepts of induction and deduction, which are at the heart of analytical thinking, were invented to find ways to reach the big picture, just like building a *puzzle*.

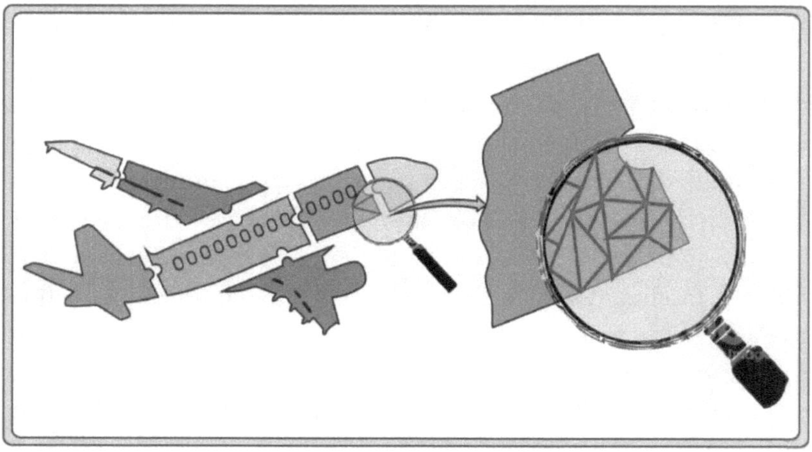

Figure 8.1: When we smash an airplane, smooth geometric shapes emerge from underneath

From primary school onwards, basic geometric shapes such as triangles, squares, rectangles, cubes and rectangular prisms entered our lives. These were easy to digest, like the small bites we would eat to reach the big picture. The fact that the width and length of these geometric shapes are clear and precise allows us to approach the big picture more accurately and meaningfully, and that is why we have learned and are learning them.

If I am asked the question "What is an integral?", I would answer that it is a mathematical function that decomposes the shapes of functions in the coordinate system into small rectangles and then combines them properly. Of course, you may encounter such definitions in the literature.

You can deduct from this answer that the integral can be used to take apart and put together a car, an airplane, a tank; to take apart and put together a computer, a phone, or to take apart and put together any event, any issue, any work. But do not perceive this breaking down and putting together in physical terms! The integral is concerned with breaking down and combining the functional relationship behind them. The main thing in integral learning is to internalize this basic knowledge and the fact that it is nothing more than breaking down the problems at hand into small meaningful pieces and then putting them together.

In all educational curricula around the world, the integral is taught through area and volume calculus. The reason why the integral is explained in terms of area or volume is related to the shapes produced by the functions we draw in the coordinate system.

Although the physical meaning of the integral is completely different, it is very appropriate to explain it in terms of its mathematical meaning, area and volume calculation. With the integral you find the area and volume of geometric shapes that you transfer to the coordinate system and whose units do not matter. This is why the integral is explained in terms of area and volume. As you will see more clearly later, this area and volume can give you the amount of matter, or it can give you the force or pressure.

Area is two-dimensional and volume is three-dimensional. I explain in more detail what is meant by multidimensional space in the matrix topic, but I just want to remind you that the increase in the number and degree of variables has paved the way for the integral to be used for much more dimensional concepts. Normally, it is sufficient for us to perceive the integral in terms of area and volume. But the integral is used to solve problems in much more than three-dimensional space. So, don't think that the integral is only used to calculate area and volume in the mathematical sense!

You cannot take the integral from the world of mathematics to the real world unless you emphasize that the shapes in the coordinate system have completely different meanings. I am trying to explain it as the world of formulas, but read it as the world of functions. Then you can see the integral in all multiplied and divided relations. Force is equal to the product of pressure and surface area, isn't it? This multiplication relationship opens a door to the integral. In the same way, the product of force and path equals work. In the same way, you have to open a path through that to the integral. Then you will see more clearly how useful this most valuable subject of mathematics is in this life.

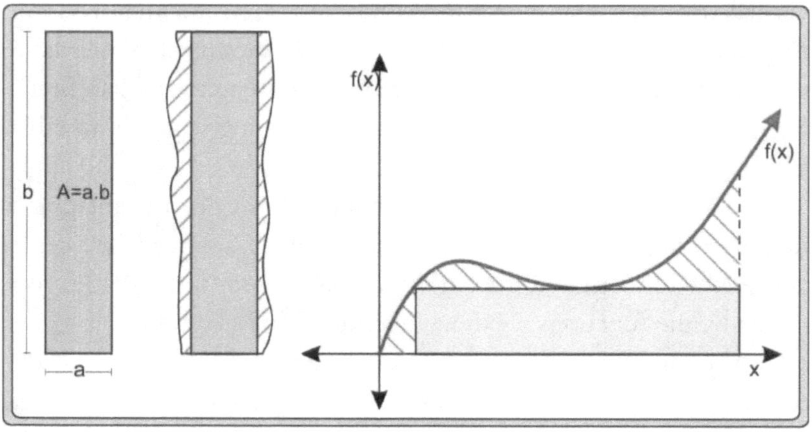

Figure 8.2: Thanks to smooth geometries, trapezoidal pictures are made meaningful.

If you want to find the area of a shape on the coordinate system, if the shape is a rectangle, you just multiply the length on the x-axis by the length on the y-axis. Problems like this are very easy to calculate. But the shapes that real problems yield are often not as smooth as rectangles. The integral is a mathematical function that arises from finding the area and volume of curved shapes.

To internalize the subject a little more, let's first estimate the area of the shapes on the next page. When I was in my university exam, there was a question where I had to calculate the integral, first I drew the graph of the function, then I roughly estimated the area of the resulting shape, I marked the closest one in the options and the result was great. Just remember that the integral does not guess, it tells you the exact and precise result. I would like to emphasize that when integrating, you need to make sure that you know roughly whether you have made a calculation error or not. People dealing with real engineering problems need to be able to roughly predict the outcome of a problem. This is like estimating the height of a person. For example, when I say how tall I am, it is as absurd to say that I am 150,000 centimeters tall as it is to say that I am 150,000 centimeters tall, and you need to know approximately the pressure you exert on the ground. In short, if you don't involve your sensory organs in the problems, you can't make a proper connection between this life and mathematics.

You may also ask, "Why do we multiply the side lengths to find the area?" Let's explain this a bit. Take a look at the rectangle below: let the short side be 3 cm and the long side 4 cm. Let's divide the rectangle into quadrilateral regions with 3 squares with a side length of 1 cm from right to left and 4 squares similarly on the next page. Let's move this to the coordinate system and then count how many squares are inside the rectangular region.

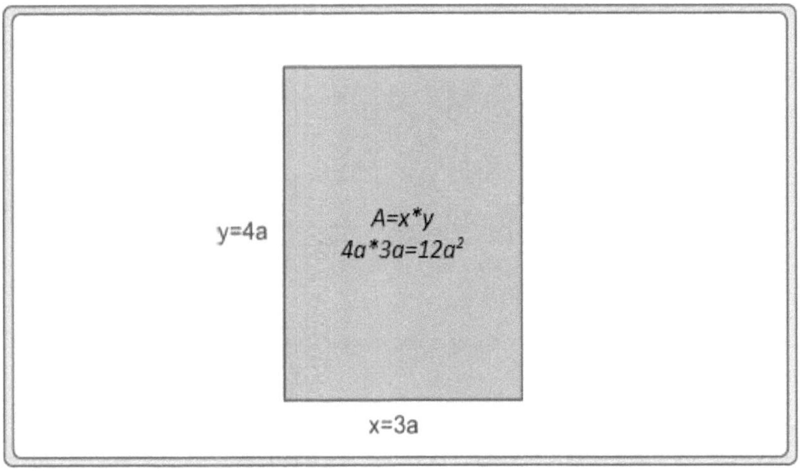

Figure 8.3: A simple field representation.

You can see that the length from right to left is 3 cm and from top to bottom is 4 cm, right? Multiply these by each other and you get the total number 12. Now count the squares; you will see 12 squares with an area of 1 cm^2 . Of course, you know that the area of a rectangle is defined as the product of the side lengths. Keep in mind that multiplication and addition have the same origin and multiplication is used as an abbreviation for addition! You need to remember these once again when explaining the integral. When you look at the area with this eye, you can easily see that the area of the rectangle above is 12 cm^2.

As you can see, the integral is taught to find the area and volume of shapes. For years after I learned this information, I thought that the integral was taught only to calculate strange shapes whose area and volume were not known. If someone had explained to me how it relates to this life, I would have realized that this definition was very inadequate and I would have understood the value of the integral at that time.

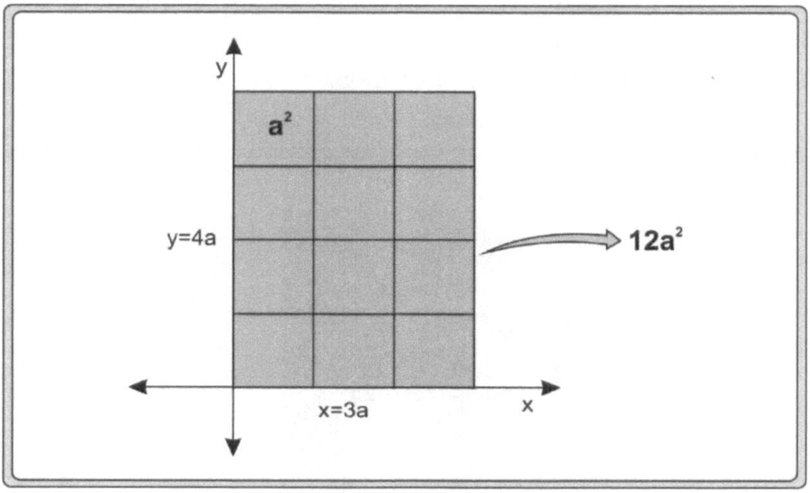

Figure 8.4: Squares that make up the area of a rectangle.

In fact, if you know what the variables in your function mean, you know what the area or volume of the shapes you get from them correspond to in physical terms. So, the physical meaning of the integral can be very different from the mathematical meaning. Of course, the connection of the integral to this life, and I am including the social sciences, is hidden in its physical meaning.

You may remember the equations of motion from physics lessons; we define a path as the product of speed and time. For example, if you are asked the question "How many kilometers does a car traveling at 60 km/h travel in 2 hours?", you immediately multiply the speed and time and find the total distance traveled, 120 km. Here I wanted you to see in advance that the path to integration passes through multiplication.

The integral of speed over time gives you the distance traveled. It will be physically absurd, but I will say something mathematically very logical; the integral of time with respect to speed will also give you the path. Because just as the result does not change when you change the factors in multiplication, the result does not change here either.

SI Base Unit	
Fundamental Quantities	SI
Length	m
Mass	kg
Time	s
Electric current	A
Temperature	K
Luminous intensity	Cd
Amount of substance	mol

SI Base Unit	
Derived Quantities	SI
Force	Kgm^{s-2}
Velocity	ms^{-1}
Pressure	$Kgm^{-1}s^{-2}$
Energy	$Kgm^{2}s^{-2}$
Acceleration	ms^{-2}
Resistance	$Kgm^{2}A^{-2}s^{-3}$
Power	$Kgm^{-2}s^{-3}$

Figure 8.5: Mathematical concepts have been enriched to understand what these quantities tell us.

Now let's talk a little bit about physics. As you know, we define physical phenomena with seven fundamental quantities. These are mass, length, time, current strength, temperature, light intensity and amount of matter. If we give a few examples of quantities derived from these 7 fundamental quantities; we can count quantities such as force, speed, pressure, energy, acceleration, resistance and power.

In fact, we can actually increase this derivation. When you include dozens of scientific fields such as chemistry, biology, medicine, geography, aerodynamics, thermodynamics, finance, etc., you can imagine how the number of derived quantities can increase, can't you? It's not hard to imagine, I just listed them, you can come across many formulas where 2 or 3 of them come together.

Now think of the combination of 7 with 2s and 3s, that is, think of formulas where two and three come together. You can also imagine that if you consider their repetitive use, i.e. when the square of the length is the area and the cube is the volume, the number will be much, much higher.

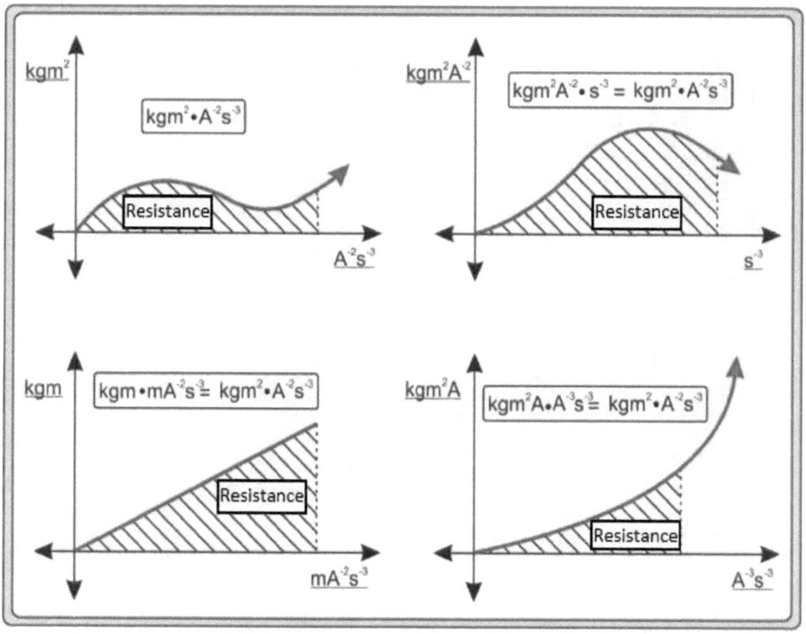

Figure 8.6: If you pay attention to the units, you will see the integral

In fact, we have an alphabet of 7 letters and you decide how many words and sentences can be made out of it. You will need the integral to find their areas and volumes in the coordinate system. Then you will understand the value of the integral better.

Using the alphabet of 7 letters you can derive many problems, you can write many functions using them. Look at the table, ms^{-2} is defined for acceleration. Multiply this by time (s) and you get the velocity (ms^{-1}). Here it is Acceleration -Time. I gave an example where you can get the velocity from the graph using the integral.

Pay attention to the units in the table, you can see that the product of mass (kg) and acceleration (ms^{-2}) gives the force (Kgms). Mass-Acceleration If you draw the graph, you can find the force here by integral calculation. You can also find the force with the graph of pressure (Kgm s^{-1-2}) and area (m^2), right? From this table, I want you to see that the product of force ($Kgms^{-2}$) and length (m) gives energy (Kgm s). $^{2-2}$

If you remember from high school and university electricity courses, voltage is equal to the product of current and resistance. Now I want you to see that when you want to find the voltage, you have to take the integral of the current with respect to the resistance or the resistance with respect to the current. You may say, "If we find the voltage by multiplication, why the integral?"

When you see the squiggly shapes of graphs, you immediately realize that life is not so rosy, that simple multiplication does not lead to immediate results. You always overcome these shapes with integration. The mystery of life is hidden in these graphs, remember that! I am trying to tell you that the integral is a mathematical operation that you can relate to every event in life. If you look carefully, you can see that every scientific subject, no matter what the subject, has a path to the integral.

In addition to the sub-branches of physics, you may come across hundreds of thousands of derived functional expressions for use in other branches of science such as chemistry, biology, finance. Imagine the number of graphs you can draw from these, and you will realize once again how useful the integral is.

Now enough of this information, let's move on to our main question: Why is the integral of x, $x^2/2$? Why $x^2/2$ and not $x^2/5$ or

$x^3/3$? If the integral is a sum, why can't we immediately see this result in this sum and get lost in formulas and rules?

In fact, this question has a critical importance for me in the emergence of this book. Because whoever I asked this question to around me, I could not get a satisfactory answer. Most of them could not even comment, some of them gave answers with memorized stereotypes, and many of the people I asked were graduates of very good universities. Those who gave the most rational answers said that it came from the field, but they could not explain how it came about. In short, no one knew why the integral of x is $x^2/2$ and no one who knew could explain it.

To me, this was a very good picture of our education system. No one knows why they are learning the subject they are learning, why the subject is being taught, and they graduate from schools by memorizing the rules they have been told, in other words, thinking that they have only learned. In fact, neither the learner nor the teacher knows the wisdom and purpose of teaching these things.

Not only pre-university education but also university education is based on memorization. This is why our universities do not even make it to the top hundred in world rankings and why most of them are labeled as high schools.

Now let's go back to our question; to find the answer to why the integral of x is $x^2/2$, let's first draw a graph of $f(x)=x$. Then fill the area between the function and the x-axis with small thin rectangles.

First of all, I want you to look at the figures and see that the thinner the widths of these rectangles, that is, the shorter the short sides on the x-axis, the sum of the areas of these rectangles will lead us to the true result.

Another point I want to emphasize here is the fact that the area of the rectangle is now a line segment. Yes, the area now appears as a line. This phenomenon is the most appropriate definition for the spirit of Linear Algebra. No matter what the problem is, turning areas, volumes, curves into lines and going to conclusions with them, that is the spirit of mathematics!

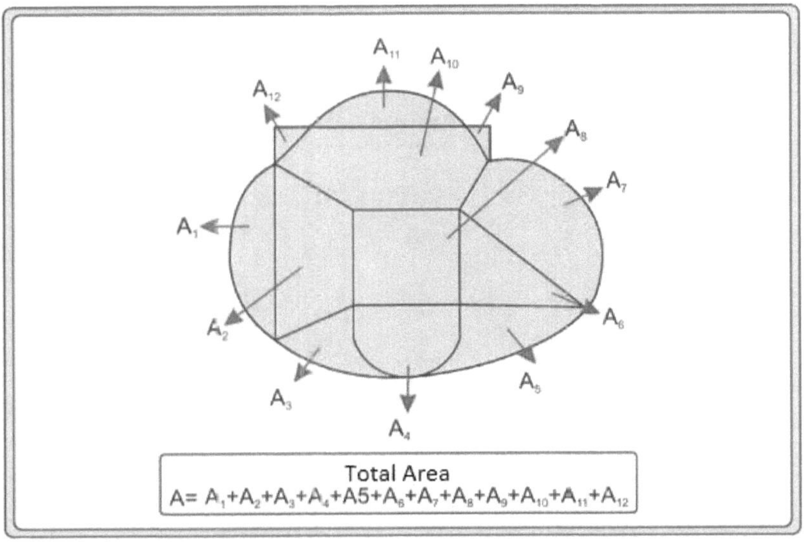

Figure 8.7: We use regular geometric shapes to get to the main image.

After all, if the integral is summation, it seems most logical to divide the shape into rectangles in order to make this summation. This is similar to summing shapes with complex geometry by dividing them into triangles, rectangles and squares, whose areas can be easily calculated, as you can see above.

If you have the question "Why do we divide the area under the function into rectangles?", the answer is actually simple. We divide the shape into lines, like spaghetti sticks, and these lines look like rectangles by the nature of their shape. If your rectangles are very, very thin, that is, if you fill the shape with rectangles of infinitesimal thickness, you can be sure that their sum will give you the exact result.

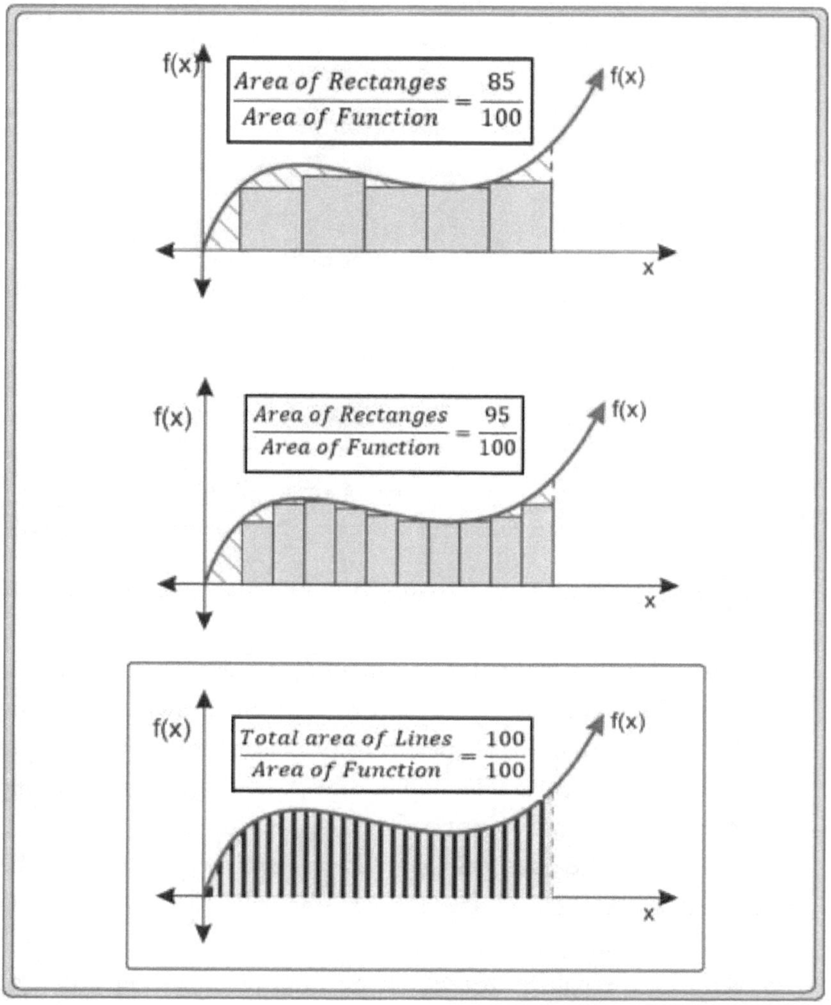

Figure 8.8: A simple field representation

Look carefully at the shapes above and you will see what I mean. Try to cover the shape with rectangles of different widths. You can see how the area of the rectangles covering the main shape increases as you order their widths from thicker to thinner, right? With the rectangles with the smallest widths - you can call them lines - you can now clearly see from these shapes that the area is completely covered.

$f(x) = x$ You can now say that in order to put an infinite number of rectangles inside the triangular region resulting from the function, the thickness must be the size of a line segment. You are now sure that if the thickness of the rectangles, i.e. the short sides of the rectangles, goes infinitesimally small, i.e. if these rectangles are made up only of lines, they will fill the triangle that gives the function its shape without leaving any gaps, right?

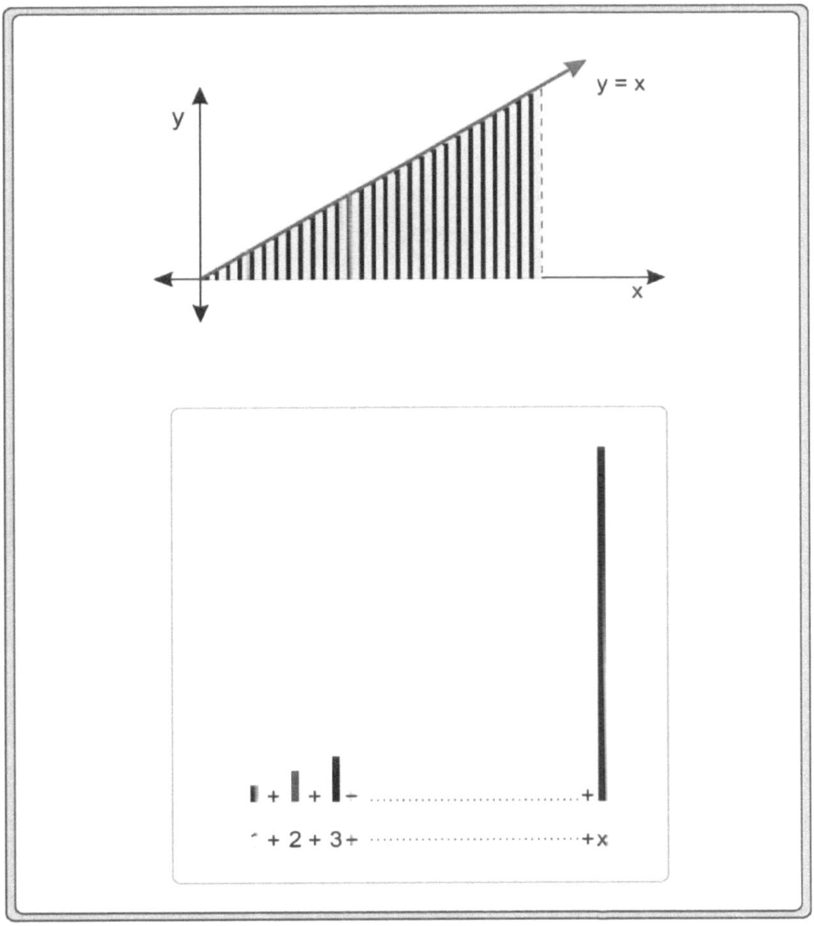

Figure 8.9: When you add the area of the lines one by one, you reach the target.

Now we are going to add to this infinite number of near-zero thin rectangles x Let's say. x What would you need to find the area of the main shape by summing the area of the rectangles? Let's change the question and ask: How do we find the total area of rectangles ordered from 1 to x? Of course, the easiest way is to use the Gaussian method. Yes, I'll let you in on a secret; on the way to integral calculus, we formulate the process of adding the areas of rectangles using the Gaussian method.

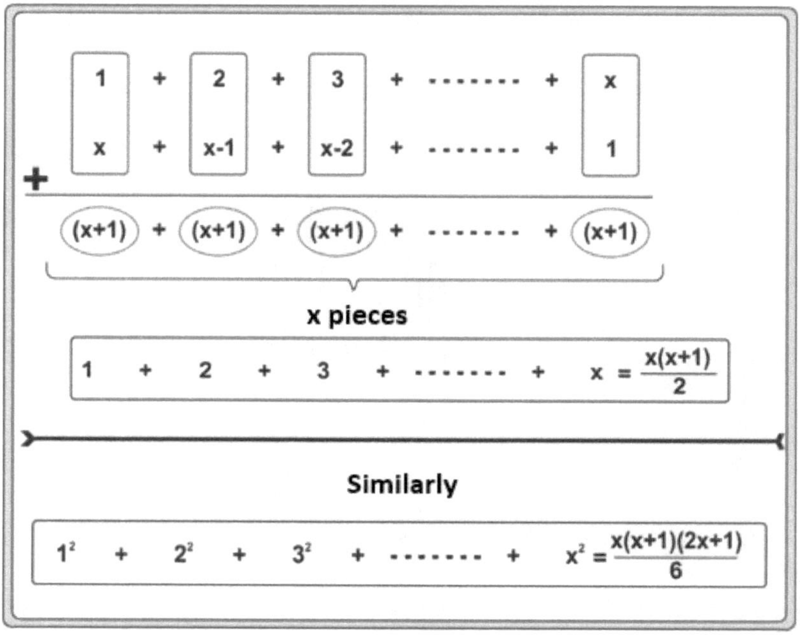

Figure 8.10: Gauss Method.

When the integral is explained, these details are explained implicitly, not underlined and emphasized. Therefore, we cannot see the expressions formulated with the Gauss method in the integral. And those who do, never tell anyone about it. This is actually not a very mysterious information, but when I wrote this book, I aimed to make it like four operations. After years, I realized how valuable it is to see a simple sum in the integral while learning that subject and I wanted to share it with you.

As you may remember from integral courses, the long side of each rectangle is defined by the function f(x) and the short side is defined by its dx. The product of these, of course, gives the area of the thin rectangle. In short, the area of each thin rectangle $Ax =$

$f(x)dx$ in the form dx. Note that to find out how many dx, the short side, we use increasing counting numbers from 1 to x.

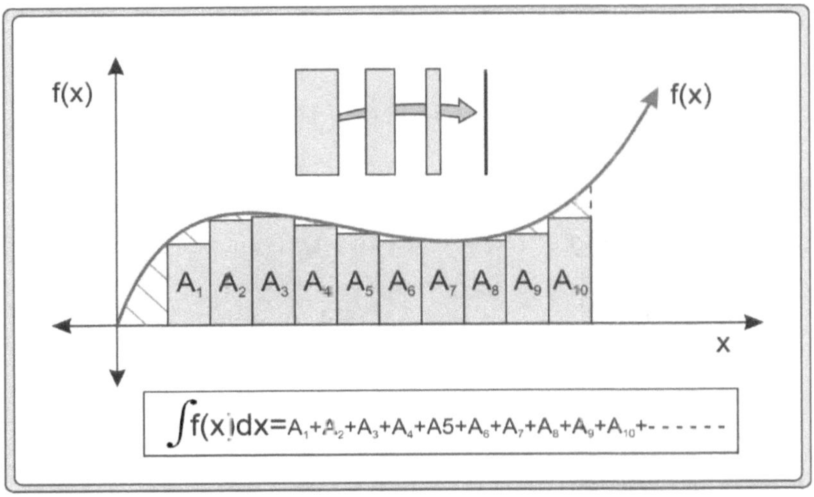

Figure 8.11: When rectangles are transformed into lines, the area of the curve approaches reality.

The other thing you need to pay attention to here is to know what the width and length of rectangles mean. First you need to see that the widths of the rectangles are always the same length (dx) and that their number corresponds to the number of terms you will use to count the rectangles. I want you to see from this graph that the sum of the areas of the rectangles is an increasing number from 1 to x, summing the number of rectangles, and that the area of each thin rectangle is represented by the output of the function f(x).

I think it goes without saying that the only things that change on the way to the area are the lengths and that the lengths consist of values such as 1, 2, 3, 4, 5... which are the output of the function y=f(x)=x that we are trying to calculate. It will not be difficult for someone who knows a little math to find the result of this sum.

Now let's digitize this sum a little more. For this, it is useful to briefly recall the Gaussian method; we found the result of the sum 1+2+3+⋯x with the equation x*(x+1)/2.

Let's remember this again with a simple example; How do we find 1+2+3+4+5= 15; If you put 1+2+3+4+5 in reverse order, you get 5+4+3+2+1 and add them one under the other;

$$\begin{array}{r} 1+2+3+4+5 \\ + \quad 5+4+3+2+1 \\ \hline 6+6+6+6+6 \end{array}$$

You get the result. In this sum, there are 5 6s as many as the number of terms and the result is 30. Since we are adding the same numbers twice, divide the result by two and you will get the actual result of 15.

This is actually not very difficult to formulate;

$$\begin{array}{llllll} & 1 + & 2 + & 3 + \cdots\cdots\cdots\cdots (x-1) + & x & (1) \\ + & x + & (x-1) + & (x-2) + \cdots\cdots\cdots\cdots 2 + & 1 & (2) \\ & (x+1) + & (x+1) + & (x+1) + (x+1) + (x+1) & (3) & \end{array}$$

Subtotal expressions 1 and 2 above and you get the 3rd sum, which is x(x+1). We can of course write this as x*(x+1). Since we have done the same sum twice here, divide the result by 2;

$1 + 2 + 3 + \cdots + x = x * (x + 1)/2$ you can easily deduce that.

You can't see the difference between 100 quadrillion plus quadrillion and one more than that, can you? So for numbers that go to infinity

$x + 1 \cong x$ acceptance is made. Remember, this is why you learn the limit!

Therefore $x * (x + 1) / 2$ of the expression $(x * x) / 2$ I mean;

x^2 /2 in the case of this equality.

Here you go $y = x'$These results tell you why the integral of x /2. 2

Now let's try to see a little more of the mystery of Gauss through another example;

$\int x^2 dx = x^3 /3$ Do you think equality is right?

$y = x^2$ why the integral $x^3/3$ is the question of what it means to be a person who is a person who is a person.

The basic joke here is that instead of x, now x^2 and the result of the sum;

$1^2 + 2^2 + 3^2 \ldots\ldots\ldots x^2$ lies in his expression.

Because the area of each thin rectangle x will be the length of the output of the function on the y-axis, and all you need to do is add these lengths together.

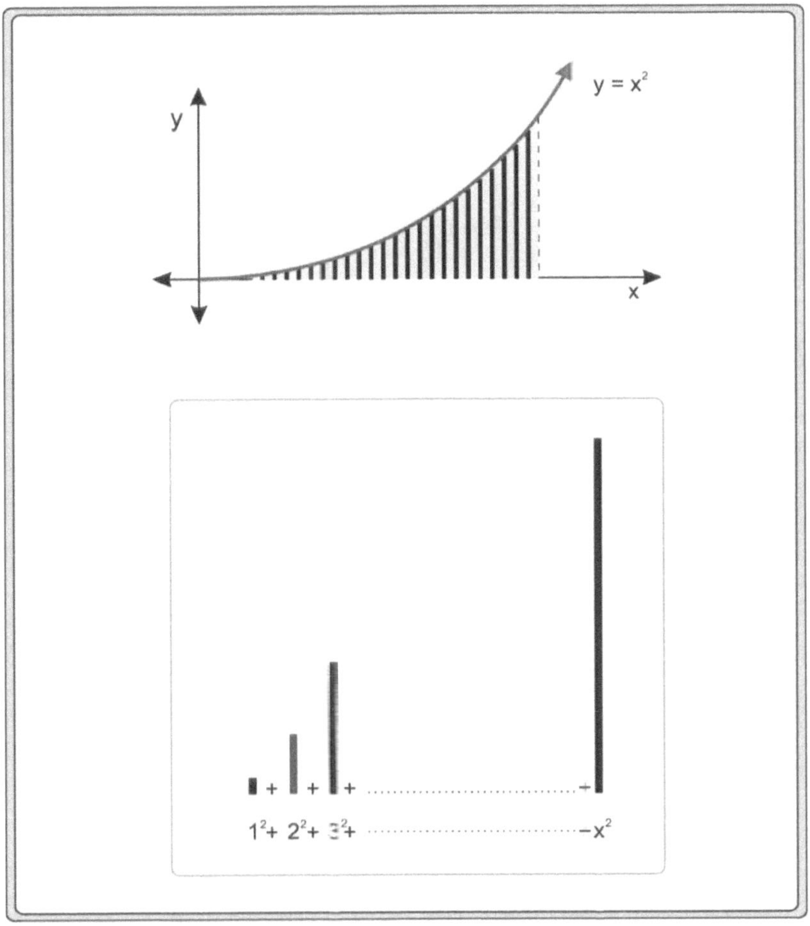

Figure 8.12: The process consists of adding the area of the lines one by one using the Gaussian method.

There are many different solutions to find the result of the sum of perfect squares, I want to show what the sum turns out to be with one of the most known methods, the polynomialization method. Let's do some math for this.

$$1^2 + 2^2 + 3^2 + \cdots ..x^2 = ax^3 + bx^2 + cx + d$$

Let's write the equation and then try to find the coefficients a, b, c and d. If you remember in solving equations, if you want to find the coefficients in an equation, you must first set the equation equal to 0.

Now $ax^3 + bx^2 + cx + d$ Using the equation = 0, we will find the coefficients a, b, c and d.

For $x = 0\,;\ d = 0$.

For $x = 1\,;$ $\quad a + b + c = 1$ (1st equation).

For $x = 2\,;$ $\quad 8a + 4b + 2c = 5$ (2nd equation).

For $x = 3\,;$ $\quad 27a + 9b + 3c = 14$ *(3rd equation)*.

You have 3 equations with 3 unknowns and we can now solve them. Let's first change it to 2 unknowns:

Multiply the first equation by 2 and subtract from the second equation and multiply by 3 and subtract from the third equation;

$$6a + 2b = 3$$

$$24a + 6b = 11$$

You arrive at the equations with two unknowns above.

When we multiply the upper equation by 3 and subtract it from the lower equation $6a = 2$ ise $a = 1/3$ we'll find it.

Substituted from the equation above;

$$2 + 2b = 3 \ ise \ b = \frac{1}{2}.$$

$a + b + c = 1 \ ise \ c = \frac{1}{6}$. we'll find it.

When we put them in the equation;

$1^2 + 2^2 + 3^2 + \cdots ..x^2 = \frac{1}{3}x^3 + \frac{1}{2}x^2 + \frac{1}{6}x$ we'll find it.

From here, if we combine the denominators and put the equation in x brackets;

$$\frac{x(2x^2 + 3x + 1)}{6}$$

form. In the equation;

$$2x^2 + 3x + 1$$

$$2x^2 + 3x + 1 = (x + 1)(2x + 1)$$

When you factorize the equation as;

$$1^2 + 2^2 + 3^2 + \cdots ..x^2 = \frac{x(x + 1)(2x + 1)}{6}$$

An infinite number of x

I've already explained that one more doesn't change the result. equation+1's and see what is left;

$x + 1 \cong x$ and $2x + 1 \cong 2x$ if you make the admission;

$x(x)(2x)/6 = x^3/3$. You get equality.

This result will give you $y = x^2$ why the integral $x^3/3$ that you're the one?

These are actually simple proofs, starting from here;

$$f(x) = sin(3x + 5), \qquad f(x) = tan(3x) + 10\,cos(2x),$$
$$f(x) = e^{2x} + 5$$

You can find out how to take the integrals of many more different functions. It is not difficult to prove them, because you can easily polynomialize these functions and then do a little bit of math to get the result. Isn't that why we learn math?

The basic maxim to keep in mind when taking integrals is the fact that after graphing functions, it is nothing more than dividing the graph into small rectangles and then summing their areas. Of course, to do this summation, you may need to master dozens of mathematical transformation expressions, from *Fourier* series to trigonometric transformation expressions.

To give you an example, I have tried to explain how the integral of 2 simple functions arises. You can of course find the integral of more complex functions by using various mathematical transformation expressions without deviating from the basic philosophy of integration. If I were the decision makers in the education system, I would give students homework to prove where the integrals of functions come from.

Now a very simple question for you:

$\int x^3\,dx$ What is the result of the integral equal to?

$$1^3 + 2^3 + 3^3 + \cdots \ldots x^3 = [x(x+1)/2]^2 \cong x^4/4$$

What does equality remind you of?

or;

$\int x^n\,dx \stackrel{?}{=} x^n/(n+1)$ Is that really true?

Integral calculus is one of the most popular solution tools for engineering problems. If you know why you take the integral in an engineering problem and where you need to integrate, you will have a better understanding of what the solution is and what it means to you.

The shapes you see in daily life, the events and changes you experience are not straightforward. Very complex lines, curves, surfaces, three-dimensional objects are all around us. I am telling you this so that you know that you have a very valuable mathematical function such as the integral in your hand in order to get out of these and to cope with them.

As I said before, the integral is not just a tool for calculating area or volume. The integral actually symbolizes a mathematical function that helps you transform a function. The concepts you call area or volume are of course also functions. If you define the functional relationships of any problem in life and want to obtain different expressions from those relationships, you will most likely need to use integration.

There is a logic in taking the integral of the area to find the volume. Just as we go to the area by joining the lines, we go to the volume by adding the areas. If you want to put these pieces together, of course you can only do it by integration. Whatever changes there are in life and you can derive functions from these changes, you can use the integral there. For example, if you want to find the distance you will travel with time-varying speed, or the amount of money you will get as a result of time-varying interest, the integral will give you the most accurate result.

Relationships, events, problems in life;

$f(x) = 2x + 3.$

When you define $f(x) = 3x^2 + 4x$, $f(x) = 5 / x^3 + 6$, you enter the analytical world. If you can make these definitions -what we call making mathematical models of the events in our lives- if you can make these models, then you will better understand why integration, derivative and function are taught.

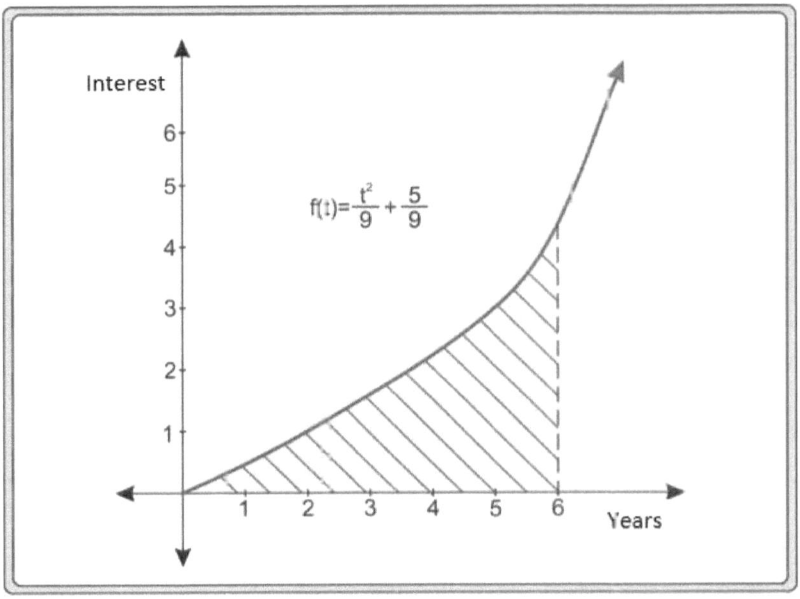

Figure 8.13: A simple example of integral calculus.

Let's do a little integral calculation to better understand this issue through interest, which is one of the important sub-headings of finance.

I have already given the general formula for interest. Interest is equal to the product of principal, interest rate and time.

Let us formulate this definition as $F=a*n*t$. On the coordinate axis, let the x-axis represent time and the y-axis represent the product of principal and interest rate. Let the principal be constant, but let the interest rate be a function of time (t) so that we get a slightly more complex graph. If you do not create a functional relationship between the dependent and independent variables, you can easily find the interest rate by multiplying the principal, interest rate and time constants. Let's add some dependence.

Let the principal a be 1000 and the interest rate n be a function of t, i.e. time, as $(t^2 + 5)/9$. Therefore, if we represent the product a*n as a function;

$a * n = f(t) = 1000 * (t^2 + 5)/9$ We can write it as.

Since the principal is fixed, we can take 1000 as 1 for now and multiply 1000 by the value we find at the end. Let's take the remaining time in interest as 6 years. Now let's draw the graph showing these together;

The area of this graph will give the total amount of interest received.

The x-axis t takes increasing values from 0 to 6, while the y-axis takes a value as a result of the function $(t^2 + 5)/9$. Here if you take the integral of the function $(t^2 + 5)/9$:

$$\int_0^6 (t^2 + 5)/9 \, dt = \frac{t^3}{27} + \frac{5t}{9}$$

You get equality.

If you say t=6, you will find 11.3.

If you multiply this value by the principal, i.e. 1000; After 6 years, the amount of interest you will receive is F= 11.300.

In almost all mathematical models of problems, the x-axis is constructed as the independent variable and the y-axis as the dependent variable. This standardized and standardized representation enables the perception, comprehension and interpretation of graphs and the solution of problems to be seen more clearly.

Usually the x-axis is represented by independent variables consisting of 7 physical quantities such as time, length, amount of matter, etc., and the y-axis is represented by dependent variables derived from the 7 quantities. But of course, on the x-axis you can also find a lot of derived quantities such as area, volume, force, pressure, etc.

I think it goes without saying that whether the axes are dependent or independent variables when integrating is completely related to the definition of the problem. I wanted to explain through the example above that when you define the mathematical model of a problem, define the functional relationships, assign one of the axes in the coordinate system as an independent variable, and the other axis has a functional relationship with this axis, then the concept of integral comes into play.

If you are going to integrate, and this also applies to differentiation, there must be a dependence relationship between the axes. If there is no dependence relationship in your function, your function becomes very simple and when you transfer these functions to the coordinate system, you see that you get lines parallel to the axes. You get rectangular regions on these lines. You can easily find their areas by multiplying the short and long sides by each other.

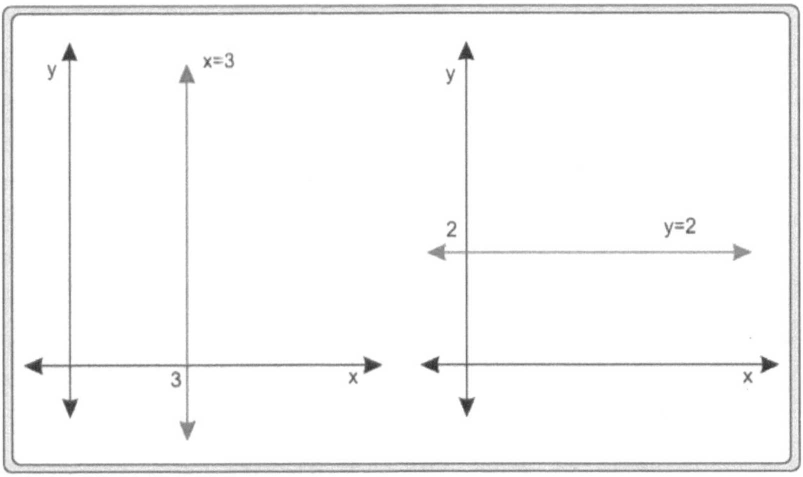

Figure 8.14: This is what an independent variable looks like. It is not affected by anything.

When you draw a curve or a line on a coordinate axis, you can observe that the curve or line is moving with reference to an axis. You can perceive the dependency relationship between the axes through these lines.

In fact, $y = f(x) = x$ or $y = f(x) = x^2$ If we talk about functions such as x, I am trying to explain that the x-axis and y-axis will be dependent on each other. Here, I wanted you to see that y is defined as the dependent variable of x while x is the independent variable, especially through the graphs of these functions.. To summarize y its value doesn't change according to its head. Every x value for a corresponding y that it has value.

You have seen graphs where the x-axis and y-axis are both independent variables and if the x-axis is integrated with respect to the y-axis, the function on the y-axis will be treated as any constant. Of course, there are such functions, but for the most part functional relationships are constructed with the pair of dependent variable and independent variable.

Now let's look at the aerodynamic problem of how to perform integral calculations to find the total pitching moment on an airplane wing.

We have moved from a financial problem to a completely different world of physics. Let's try to see here how mathematics and two different disciplines are connected.

Let our aerodynamic problem be the following; What is the total pitching moment on the wings of an airplane with a rectangular wing, a total wing area of 50 m^2 , a wing chord length of 5 m, a coefficient of restitution of 0.6, flying at 60 m/s at an altitude of 10,000 meters?

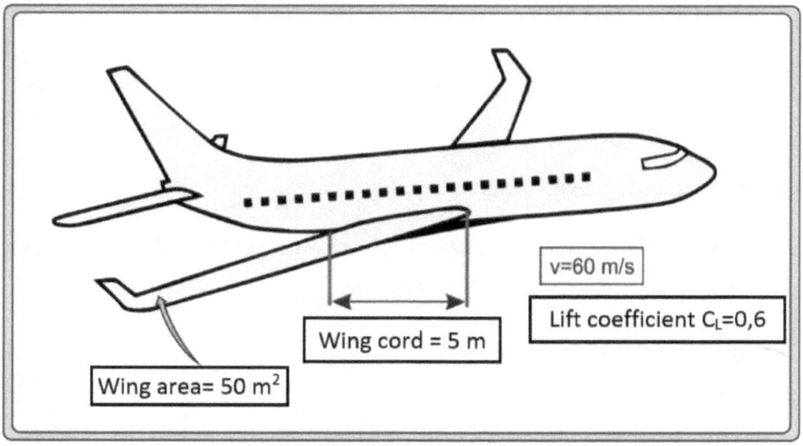

Figure 8.15: Another example to illustrate integral calculus.

Let's take a look at how this problem is solved with the basic rules of mathematics and, of course, the integral:

Now let's see what we have; let's assume that we have basic data such as the transport coefficient 0.6, which is an experimentally pre-calculated unitless coefficient according to the shape of the airfoil, the wing area 50 m^2, the speed of the aircraft relative to the ground 60 m/s and the density of the air at 10,000 m altitude 0.4 kg/m^3 .

First of all, let's remember the basic formula for pitching moment; some of you may remember the pitching formula from aerodynamics lessons. Let me write this formula here, the pitching moment of an airplane in the literature;

$M = \frac{1}{2} * C_L * \rho * V^2 * S * c$ is defined as.

Here C_L transport coefficient, ρ density of air, V the actual speed of the airplane, S the wing area, c is the cord length. In this problem, let us denote the actual speed of the airplane by V and define it as a function of the speed of the airplane relative to the ground (V_y).

When you look at the formula for momentum, we can easily say that it is technically no different from the formula for interest, just by looking at the multiplied expressions. Take another look at the world of formulas and you will see that almost all of them involve 4 operations and especially multiplication. Then you will understand more clearly how the integral is useful among them.

Of course, in order to be able to construct such problems, you need to know the physics of the subject, that is, what is what. However, this book tells you that even if you do not know the physics of the subject, you can solve such problems very easily with only your knowledge of mathematics.

Let us now look at the basic assumptions for this calculation. Since the airplane will fly at a certain altitude, let the density of the air be defined as a constant calculated according to this altitude. Likewise, let the wing area be treated as a constant. However, considering that the actual velocity, which directly affects the pressure distribution on the wing of the airplane, i.e. the bearing force, will vary along the chord, let's write this velocity as a function of the airplane's speed relative to the ground and the chord. As for the bearing coefficient, remember that it is a unitless coefficient calculated by experimental analysis that varies according to the airfoil!

Now we will find the total moment by taking the center of gravity as the origin along the chord, i.e. integrating from (-3) -(2) meters. We will write the actual speed on the wing as a function of the actual speed of the airplane and the variation along the chord, so the variable x will represent the variation of the speed along the wing chord. Now I am making up the function in my head;

$V = V_y * \sqrt{(\frac{x}{x^2+1})}$ Let it be a function of the form. $V_y =$ 60 Since m/s is $V = 60 * \sqrt{(\frac{x}{x^2+1})}$ to show the problem solution. We will use this function to show the problem solution.

$$M = \frac{1}{2} * C_L * \rho * V^2 * S * c$$

$$dM(x) = \frac{1}{2} * 0.6 * 0.4 \frac{kg}{m^3} * 60^2 \frac{m^2}{s^2} * \left(\frac{x}{x^2+1}\right) * 50m^2 * dx$$

When you multiply constant values, paying attention to the units;

½*0.6*0.6*0.4*60² *50=21,600 kgm/s^2 is found.

$dM(x) = 21.600 * \left(\frac{x}{x^2+1}\right) * dx$ is written as a function.

$\int_{-3}^{2} 21.600 * (\frac{x}{x^2+1})dx$ when you get it,

$21.600 \int_{-3}^{2} (\frac{x}{x^2+1})dx = 21.600 * ln\sqrt{(x^2+1)}$ we'll find it.

When you set limits between (-3)-(2) meters for the total pitching moment;

$$M = 21.600 * ln\sqrt{(x^2+1)}\Big|_{-3}^{2} \cong 7.484\ Nm$$

Approximate pitching moment 7.4 kNm.

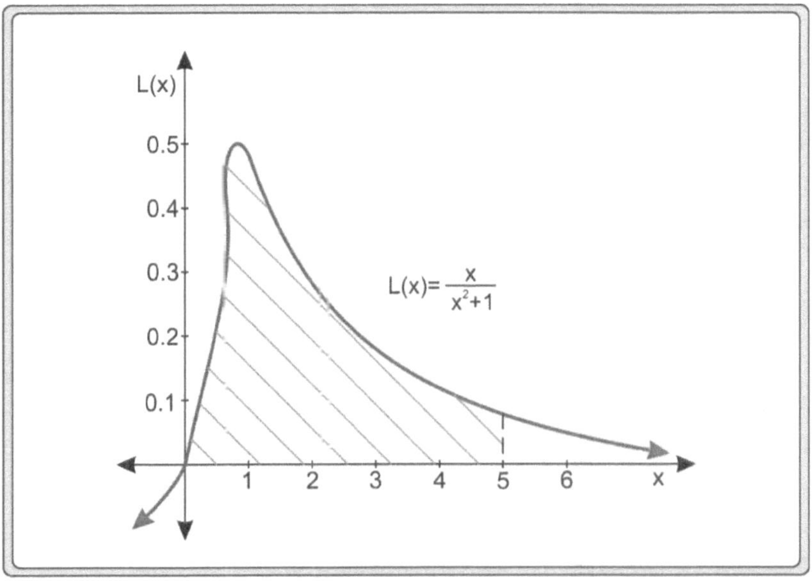

Figure 8.16: Graph of L(x) function.

As you can see, when you take the integral of this function along the wing rib, i.e. with respect to x, you get the total pitching moment. Here I would like to remind you again that the independent variables become a fixed number in the integral calculation.

The density of the air could be written as a function of altitude and temperature, and the wing area as a function of the chord. In this problem I did not write these functions to simplify the solution. Would it have been too complicated if I had? Of course not. After all, it is obvious how to find the bearing force. There is only multiplication, is it difficult to multiply functions among themselves? The important thing is to extract the main function. Once you do that, math will take care of the rest.

The solution of these problems, which come from two different disciplines, illustrates the fact that in integral calculus, once you know what you call a constant and what is a function of what, solving the problem becomes a piece of cake.

In conclusion, let us emphasize once again how important the function is when explaining the integral. When functions are transferred to the coordinate axis, I have already mentioned that the x variable on the horizontal axis often represents the independent variable and the y variable on the vertical axis often represents the dependent variable. You can consider the concept we call independent variable as an expression that changes spontaneously, without depending on anything. I can say that the time variable in particular is a good example of an independent variable. Changes in speed, acceleration or position depending on time can be encountered in continuous problems. These are examples of time dependent variables.

However, one should not make a mistake here either. The independent variable can be independent only for that event. In one problem, the independent variable can be a dependent of another expression. Therefore, in functional relationships, whether they are nested functions or inverse functions, remember that the dependent and independent variables can constantly change within themselves! The most basic principle in problem posing and solving is that you decide how to define the dependent and independent variables according to the structure of the problem.

For example, if you want to define a relationship between the weight of the airplane and the tire pressure, the weight of the airplane can be an independent variable and the tire pressure can be a dependent variable that changes with the weight of the airplane. If you define the problem in the opposite way, that is, if you want to solve a problem like "What should the weight of the airplane be when the tire pressure changes?", then the dependent and independent variables are switched. Here we have also given an example of the inverse property of functions.

No matter how many events, issues and problems you have in your life, you first need to create a mathematical model of them. After separating the unknowns into dependent and independent variables on the model and formulating them, it will be much easier to integrate the function and derive different meanings from them.

You can define a wide variety of problems in life. It depends on your imagination and your needs. For example, you can define many different problems ranging from the fuel consumption of a car and the distance you travel, the movement of the cylinders and the total power you get from the engine, the speed of a bullet from a gun and the distance the bullet will travel, the weight of the bullet and the rate of deflection in the target.

You can, of course, increase the number of such problems many times over, from calculating the maximum speed that a ship of the same weight can reach by using engines of the same power and changing only its geometry, to finding the relationship between speed and surface area for the movement of an airplane's winglet. As long as you want problems. You can see problems everywhere you look.

If you observe well, you will see that there are millions of problems waiting to be solved in daily life. In order to solve them, create your model, determine your variables, write your function and mathematics will do the rest. This is a job for those who prepare the education curriculum! They should write original, real problems, determine their dependent and independent variables, and make changes in the curriculum for functional analysis.

Now you can understand why the integral of x is $x^2/2$ or why the integral of x^2 is $x^3/3$, right? Of course, the concept of integral is a sum, and the meaning of integral with mathematical transformations was a bit hidden among formulas, letters, numbers, symbols. I tried to make it a little bit visible. You learned that mathematical expressions are derived to understand the events in our lives, and now you have the answer to why you need to integrate when you face a problem. If there are problems to be solved somewhere, you can find them by integrating with peace of mind and you can now say what the result means without fear.

Derivative

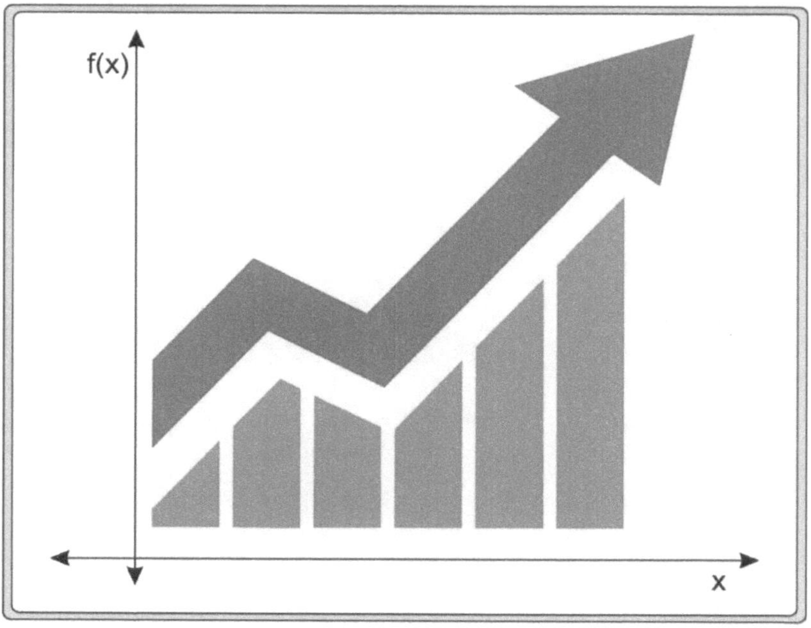

Why Do We Take Derivatives?

In mathematics courses, the concept of derivative is mentioned together with the integral and is taught as the opposite of the integral. In fact, in the current curriculum, since the derivative is taught before, the integral is taught as the inverse of the derivative. In other words, the issue of "Did the chicken come out of the egg or the egg come out of the chicken?" comes up here as well. Some of you may remember the books that explain the integral as the sum of the derivatives of the related function. In my opinion, this definition is very inadequate. With such a definition, you can never convey the logic of two of the most important subjects of mathematics. I wanted to explain the integral first and then the derivative, as I find it more accurate to teach the main picture by breaking it down.

In mathematics education, by the end of high school, students are taught the concepts of derivatives and integrals and are expected to internalize analytical thinking and demonstrate their ability to solve real problems using these concepts.

We need to understand that it is a waste of time to expect scientific works or technological products from people who think they have learned subjects by memorizing them without knowing why they are learning and without questioning anything. If we can model the education system in the light of questions such as "Why are these subjects learned and where are they used?" then we can raise truly evolved individuals who will change the balance in the world.

When people around me ask me, "Where did the derivative come from, why is it taught?" I start by explaining that the derivative emerged in order to model and analyze all the changes we encounter in life and to make meaningful inferences with mathematical relationships between changes.

I think that the life we live is in constant change, that everything undergoes change and transformation over time, that this change tells us something, and that if we can read this change correctly and model it properly, then we will be able to decipher the code of life. The only constant fact in this life is change. The derivative stands out as one of the most fundamental mathematical concepts that draws a picture of this change and helps us understand it.

Physical change of a person over the years

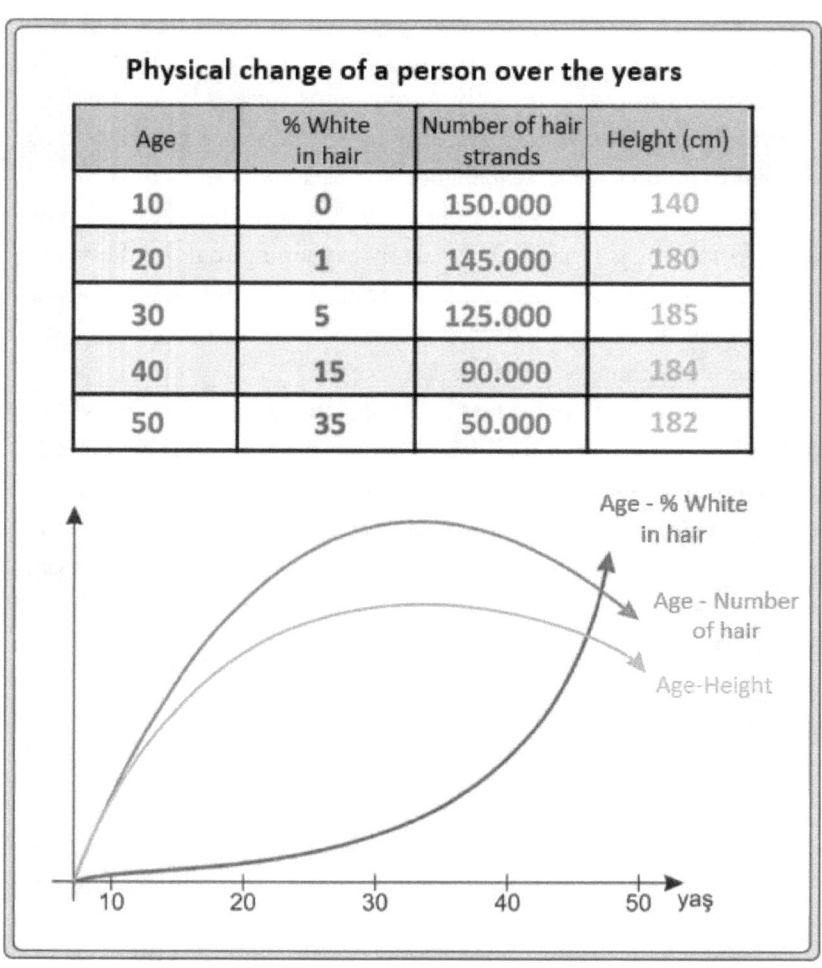

Age	% White in hair	Number of hair strands	Height (cm)
10	0	150.000	140
20	1	145.000	180
30	5	125.000	185
40	15	90.000	184
50	35	50.000	182

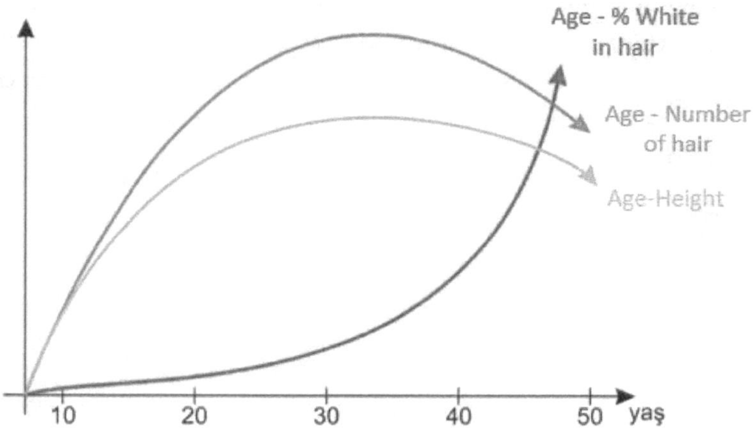

Figure 9.1: You can take the derivative of each change.

If you don't keep up with change, you will fall behind and then disappear. If the US, the winner of the Second World War, were to go to war now with the weapons it has, it would be destroyed in a few days. Technology is in a constant state of flux and we even find it difficult to keep up. Nokia, which was the leader of the cell phone industry in the 1990s, is now at the bottom because it cannot keep up with this change.

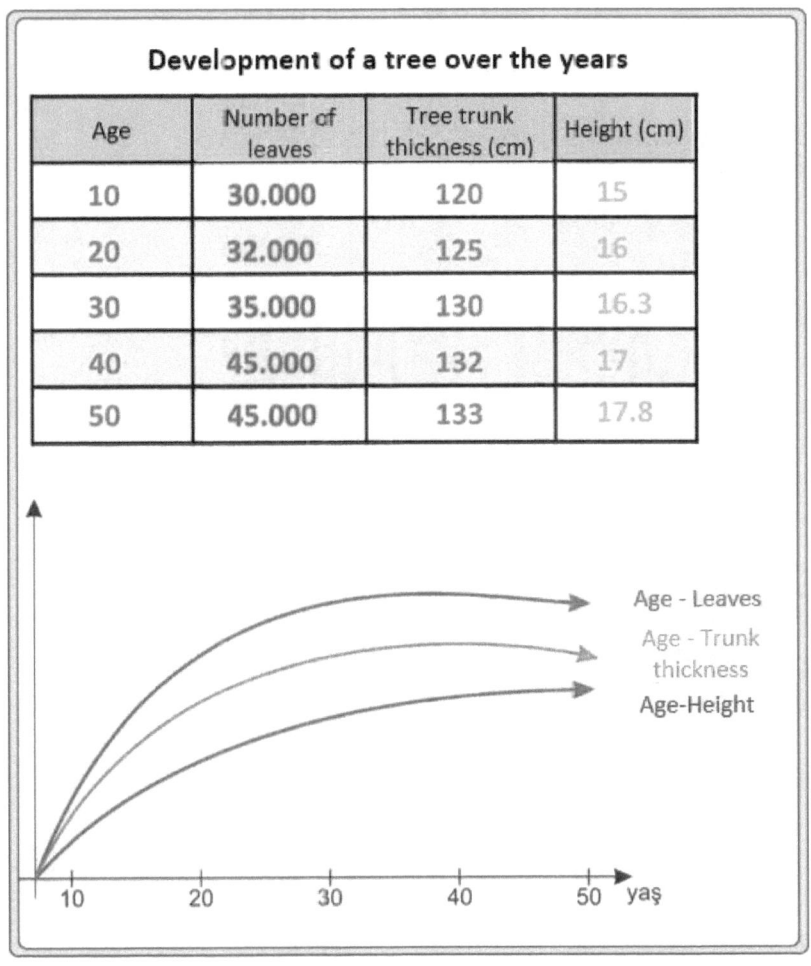

Development of a tree over the years

Age	Number of leaves	Tree trunk thickness (cm)	Height (cm)
10	30.000	120	15
20	32.000	125	16
30	35.000	130	16.3
40	45.000	132	17
50	45.000	133	17.8

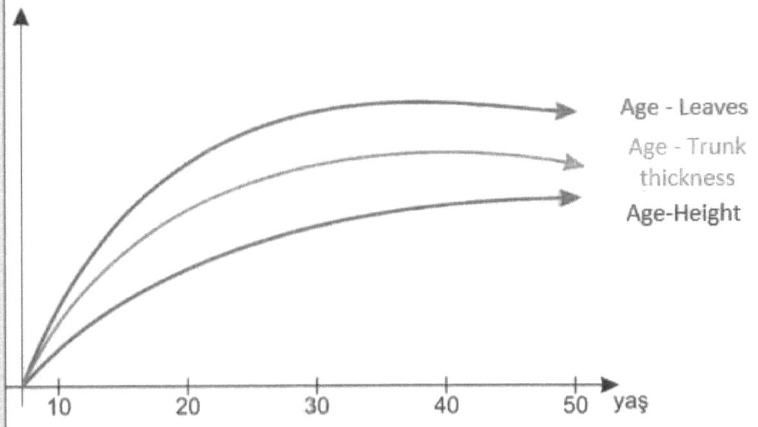

Figure 9.2: You find the rate of change with the derivative.

We see change everywhere. From events that change with time to events that change with temperature, from events that change rapidly to events that change with altitude, from changes in exchange rates and stocks to changes in inflation, we use mathematical concepts to make sense of, analyze and interpret many events.

In many parts of this book, I often talk about the concept of variables. I explain that we make our lives meaningful by analyzing how variables are relative to each other. Life is in constant change and there is a big difference between yesterday and today.

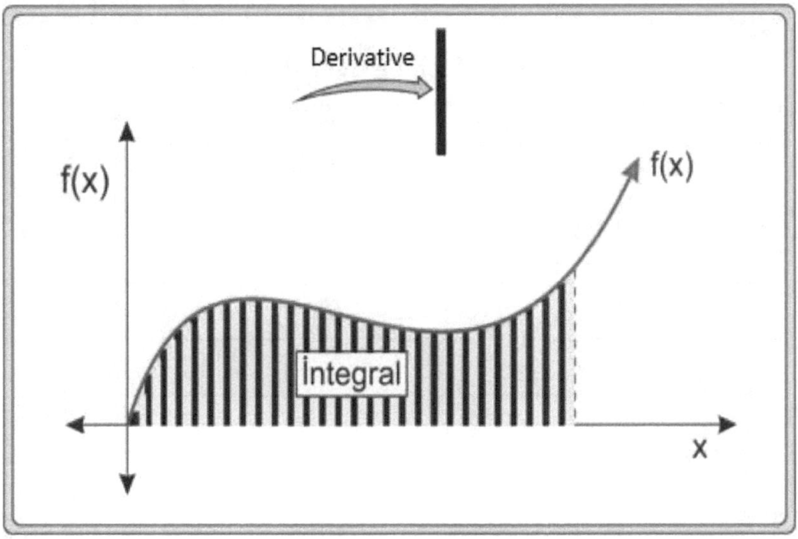

Figure 9.3: You can also understand the derivative with lines.

You may have one good day and another bad day. You don't always deal with the annual balance sheets, you want to examine the details of it. In other words, you want to analyze what happened day by day. Derivatives give you the opportunity to understand and interpret that. Learning about annual earnings can be valuable. However, if you want to go into the details of annual earnings, you may want to access more valuable information such as what you earned on which day, when your income increased and when it decreased. Here, the integral will help you to see total earnings, and the derivative will help you to know your earnings on any given day. You will understand what this means better when you finish this book.

We called the integral sum. The abbreviation for addition is multiplication. If the integral is sum, the derivative is subtraction; if the integral is multiplication, the derivative is division. On the subject of integration, I have already explained that we find the area of a function by dividing the basic shape into smaller rectangles and then adding their areas one by one. To find the area of one of the rectangles that make up the basic shape, you need to subtract the area of all the rest from the total area.

If you transfer a function to a coordinate system and deal with the area of the curve there, you need an integral; to find the line segments that make up the curve, you need a derivative. The derivative and integral are based on such a simple foundation. The mathematical meaning of their operation is nothing more than what has just been explained. When shapes and symbols are involved, it sounds like building a spacecraft, doesn't it? In fact, we call derivatives and integrals the university-graduated version of the four operations.

Now imagine a rectangle, what do you do to find its area? You multiply the sides by each other, right? The derivative is a concept that emerged to find those sides. The reason why I say, "If the integral is the area, the derivative is the edge," is hidden in this logic.

"If you ask, "Why does the derivative describe change?", I begin by saying, "First of all, think of a curve. Aren't the lengths of the perpendicular lines that make it up constantly changing? Each of these resulting lines describes change. If you say, "How do we make sense of this?", you first need to be able to clearly define the functional relationships of your problem. The mathematical meaning of these functions is given by the curves you create on the coordinate system. From these curves you get meaningful expressions through derivatives and integrals.

If we open the subject a little more with real problems in daily life; for example, iron, one of the most used materials in this life, rusts when it stays in the open air for a long time, and the rusting time varies depending on the condition of the environment in which the iron is located. Buildings age over time and then collapse at the slightest inconvenience. People also age over time, and factories can grow and shrink over time. Such changes are indispensable facts of life. The derivative is a mathematical function designed to understand, analyze and interpret such changes.

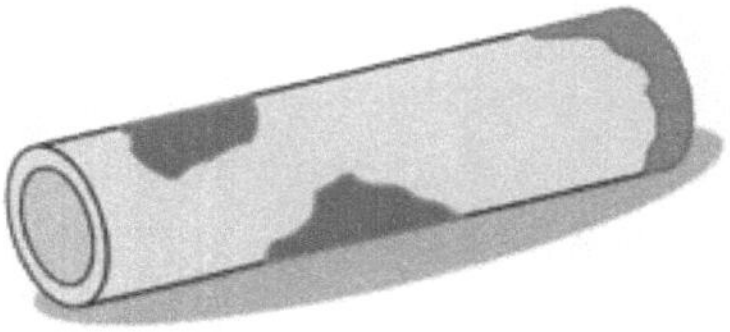

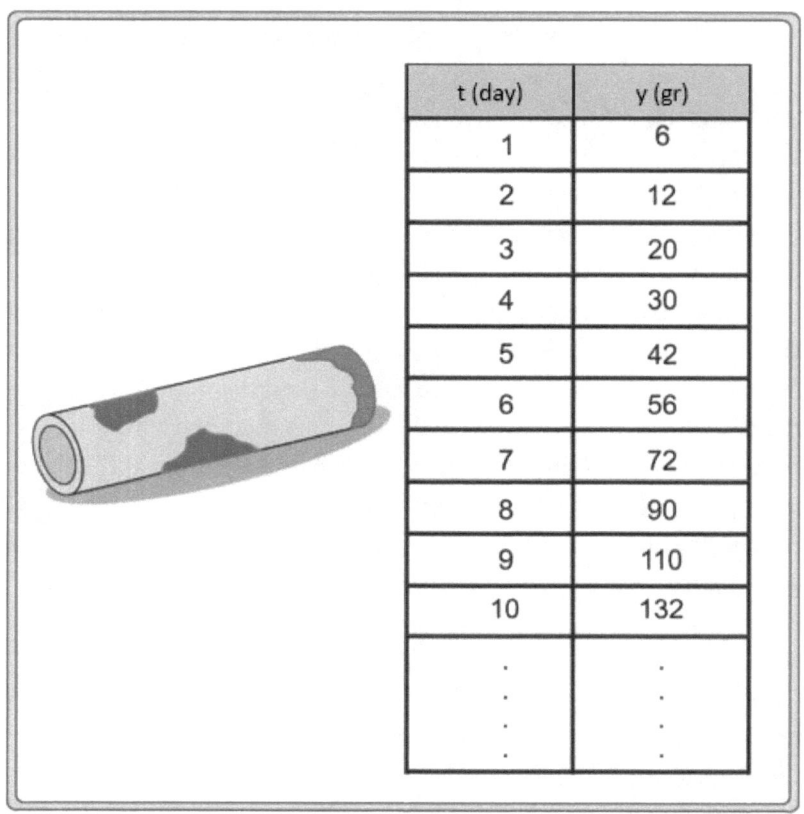

t (day)	y (gr)
1	6
2	12
3	20
4	30
5	42
6	56
7	72
8	90
9	110
10	132
⋮	⋮

Figure 9.4: A simple table of changes used to illustrate the derivative.

Now let's continue to explain what the derivative tells us through a simple example. First of all, we will need numbers to understand the subject. Let's write a problem and try to obtain a function with concrete, tangible expressions. Let the amount of rust accumulated on 1000 tons of iron left in a humid area be 132 g after 10 days, and after 100 days, let the amount of rust accumulated on it increase very much, for example 10.302 g. Now let's make continuous measurements from day 1 to day 100 and we will have a data set showing the amount of rust on each day.

Let us remind you once again that almost all formulas are special functions derived from the data set you have obtained as a result of the experimental measurement. Now, if you record the values you obtained as a result of the measurement as time (t) and quantity (y) and then connect them with lines, you will transfer the graph of your experimental data set to the coordinate system.

Of course, there are various ways to find the function from the graph. Take a look at how to find the function itself from the graphs. It takes some time, but you can find the function even by hand calculation. With the introduction of the computer in our lives, computational processes have become much faster, but the logic of solving the problem has not changed from a technical point of view. Therefore, you can also find the function from the data using computer software packages.

Let's assume that from the graph resulting from your data set, we find a function like $y(t) = t^2 + 3t + 2$, which represents the total amount of passes. Once we have the function, we can now play with it like a ball. While your experimental results give you a very small picture, with this function you have the opportunity to see the whole picture. You can now predict what that iron will look like in 10 years, or how much rust will accumulate on it on any given day.

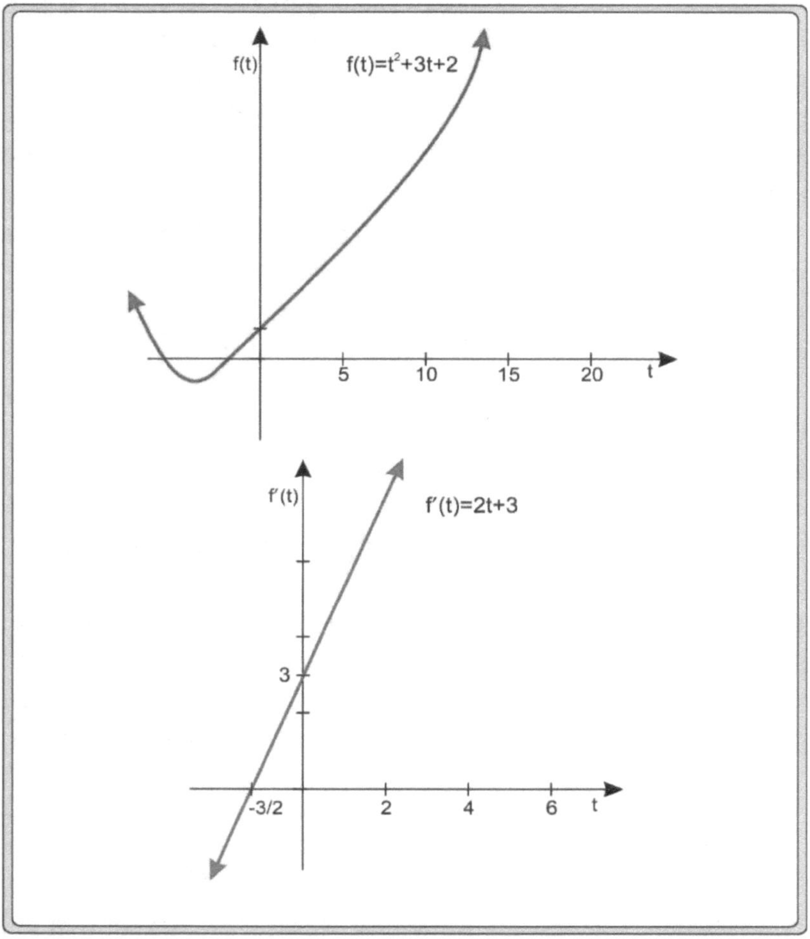

Figure 9.5: A simple example function for understanding the derivative.

I will explain the mathematical logic of taking the derivative later. Now I want to show you what the derivative tells you concretely by taking the derivative of the function $y(t) = t^2 + 3t + 2$. When you take the derivative of this function, your corrosion rate function;

$y'(t) = 2t + 3$ so that $y'(10) = 23$ for $t = 10$,

For $t = 25$, $y'(25) = 78$.

"Why do we call the derivative of a function the speed of rust?" you may ask. The most basic logic lies in the triad of position, velocity and acceleration. This triad is one of the most basic tools for understanding the integral and derivative. The basic teachings of physics are explained through this triad. They are transformed into each other by taking their derivatives and integrals, thus showing how and where derivatives and integrals are used. But it is forgotten that these are actually examples. We only see the derivative and integral in them and most of the time we cannot get out of this picture. In other words, we cannot take the derivative and integral much beyond the triad of path, velocity and acceleration, which we encounter a lot in daily life.

In fact, these three are concepts that have been specially chosen to grasp and perceive these issues. Because within these expressions lies an independent variable called time, which is at the center of life. We learn these concepts as a function of this variable. If you do not extend them, derivatives and integrals will remain very limited and will not fulfill their purpose. If you want, take a look at the world of formulas; try to perceive how expressions change and what they become by taking derivatives and integrals. Then you will better understand what integral and derivative mean.

Now let's step out of this picture; for this example, if you perceive the total amount of rust as the path, then you can call the daily amount of this as the speed of rust. As you can see, the triad of position, velocity and acceleration allows you to better understand basic concepts such as in this example.

If you take the derivative of the main function, you can see, for example, that the corrosion rate at day 25 is much higher than the corrosion rate at day 10. If you ask, "What good are these values?" I can tell you that you need to take measures to reduce the corrosion rate of the iron you have, or you can get data that shows when and how much load the iron can withstand. This rate of corrosion gives you access to facts. For example, on the 25th day you may find out that it is raining. Or you can infer that the weather is getting worse as the days go by.

If you take measures against rust, of course your rust rate function will change. Now let's assume that your function $y(t) = t^2 / 3 + 2t + 5$, which shows your new amount of rust after the measure you have taken. If you take the derivative of this, your rusting rate function is $y'(t) = 2t / 3 + 2$. For $t = 10$, the rusting rate $y' = 8.66$ and for $t = 25$, the rusting rate $y' = 18.33$. This means that while the amount of rust accumulated on day 10 was 8.66 g, the amount of rust accumulated on the iron on day 25 was 18.33 g. Whether you coat the iron to reduce rust or reduce the humidity in the environment is up to you. If you have measurable values with the measures you take, you can achieve the desired goal with cost-effective and effective solutions within the limits of analytical thinking.

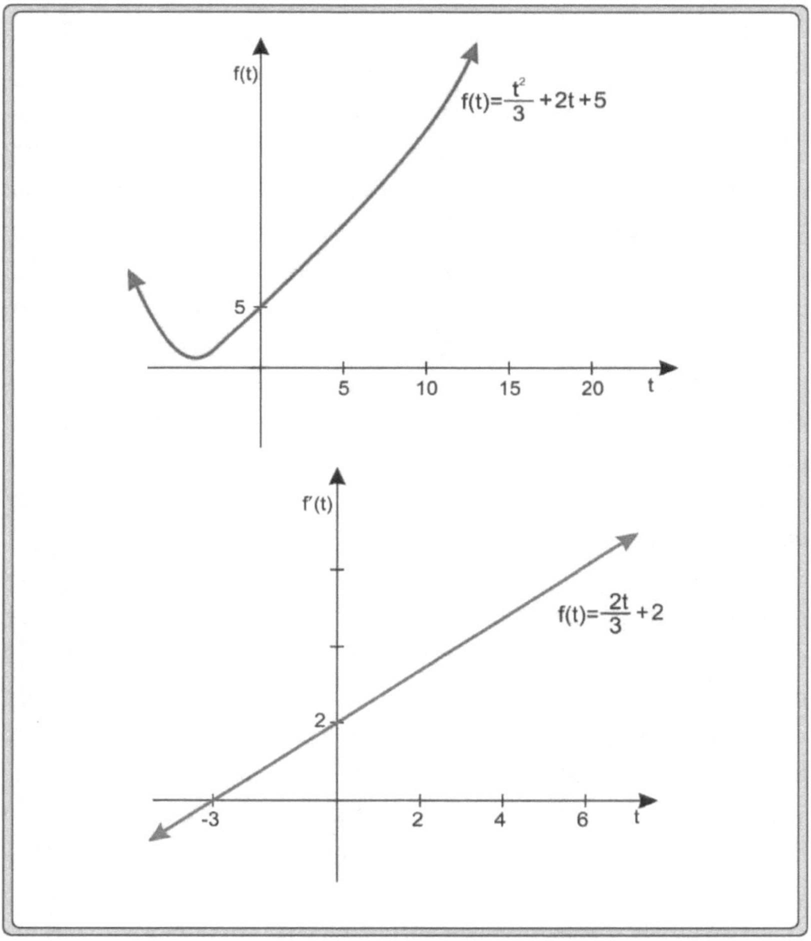

Figure 9.6: Another example function for understanding the derivative.

In order to use mathematical functions such as derivatives and integrals in real problems, you must first turn every event and every problem in life into a problem and make a mathematical model of it. Then, after finding and extracting the functions from the problem, if you can transfer them to the coordinate system, you will have done most of the work. So when you define your problem with variables like x and y, and then transfer them to the coordinate system, you can get the meaningful shapes, expressions and results you need.

In order to answer where and why mathematical functions such as derivatives and integrals are used in engineering problems, we need to know their mathematical meaning. You need to know the answers to basic questions like "Why did this become that?".

I have already explained the mathematical meaning of the integral. Now let's start explaining the mathematical meaning of the derivative through the derivative of $y = x^2$ why $y' = 2x$ and the derivative of $y = x^3$ why $y' = 3x^2$. If we can learn where the derivative of these two functions comes from, we can now approach more meaningfully where the derivative of any function comes from.

When we build mathematical models of real problems, we use symbols and expressions like x and y, and we usually call x the independent variable and y the dependent variable depending on x. As I mentioned while explaining the integral, an independent variable is a variable that changes spontaneously without depending on anything in the problem you are solving. A variable that is an independent variable in one problem can, of course, be the dependent variable in another problem. The dependent variable is the variable that changes with the change of the independent variables. Remember. Dependent or independent variables are concepts that arise from the way you define the problem.

The derivative is the name of a state that shows the effect of infinitesimal changes in any value of a function. Now let's explain what the derivative becomes with small changes in a function without using formulas. If you ask "How do we find this?", let's first determine a fixed point on the function and look at the effect of very, very small changes there.

Now you can easily find that when the value of x increases from 1 to 1.1 in the function $y=f(x) = x^2$, the value of y increases from 1 to 1.21. Without deviating from the notations of derivatives, let's call the very small amount of increase on the x-axis dx and on the y-axis dy.

For x value 1

$$dx = 0{,}1$$

increase on the y-axis when the amount increases.

$$dy = 0{,}21$$

you'll see.

Let's make this increase a little smaller, i.e. let x go from 1 to 1.02. This time the y value will increase to 1.0404. Amounts of increase $dx = 0{,}02$ ve $dy = 0.0404$.

When you proportion these $\frac{dy}{dx} = 0{,}0404/0.02 = 2.02$. As you can see, the amount of increase is very very small, so we can say that the result goes towards 2.

Note for now that when we make an infinitesimal change at the point 1 for the function $f(x) = x^2$, we get the number 2.

Similarly, let's take a look for point 3; if the x value goes from 3 to 3.1, the y value will go from 9 to 9.61.

Difference $9{,}61 - 9 = 0{,}61$ ve $3{,}1 - 3 = 0.1$ will be. When you proportion these $\frac{dy}{dx} = 0{,}61/0.1 = 6.1$.

If you want, make the increase a little smaller and you will see the result more clearly. When the x value increases from 3 to 3,001, the y value will be 9,006001. First find the differences, dy=9,006001-9=0,006001, dx=3,001-3=0,001 and then divide them by each other and you can see that dy/dx=0,006001/0,001 is almost 6. Here, when you make an infinitesimal change at the point 3 in the function $f(x) = x^2$ you get the number 6.

These have a logic. We were doing addition in the integral, we called multiplication the short way of addition. Here we do the opposite and we do division to see the change. Now let's make this meaningful;

Let's call the small change in x x+h. h here represents an infinitesimal change. If your function $f(x) = x^2$, then $f(x+h) = (x+h)^2$. The process we do is just proportioning them. As you know, in the literature, the derivative is defined by the following equation.

$$f'(x) = \lim_{h \to 0} \frac{f(x+h) - f(x)}{h}$$

This basic notation summarizes everything.

Now we are going to open the derivative equation for the function $f(x) = x^2$. First of all

$(x + h) = x^2 + 2xh + h^2$ note the equality.

$f'(x) = \lim_{h \to 0} \frac{x^2+2xh+h^2-x^2}{h}$ You can write 's. When you make the necessary simplification with the assumption h=0, using the concepts from the limit;

$f'(x) = \lim_{h \to 0} \frac{2xh+h^2}{h} = 2x$ you can easily see that.

Now this equality tells you that the derivative of the function $f(x) = x^2$ is $f'(x) = 2x$. The rate of change at x value 1 will be 2, and the rate of change at x value 3 will be 6.

In short, the functional expression $f'(x) = 2x$ gives you a result that you can interpret in general terms. This is, of course, also a function and you can graph it. Now, when you visualize these expressions, you can understand and interpret what and how you want to achieve much more easily.

$f(x)$ = x Why is the derivative of ^{3} $f'(x) = 3x^2$? Now let's illustrate this question in a similar way. First, let's try to find it by minimizing the increase around a point without using a formula;

When x goes from 1 to 1.1 $f(x)$ For the value 1, you get 1.331 from 1. Proportion the differences, 0.331/0.1 results in 3.31. Make the increase even smaller. When it goes from 1 to 1.01 $f(x)$ From a value of 1, you get 1.030311. Proportion the difference again. 0.030301/0.01 results in 3.0301. As you can see, the result goes towards 3.

Now let's choose a different x value, when x goes from 3 to 3.1 $f(x)$ increases from 27 to 29,791. Proportion the difference again. From 2.791/0.1 the result goes to 27.91 and you will see that the ratio is rounded up to 27 when you decrease the increment amount.

Now $f(x) = x^3$ why the derivative $f'(x) = 3x^2$ Let's find it with basic differentiation logic:

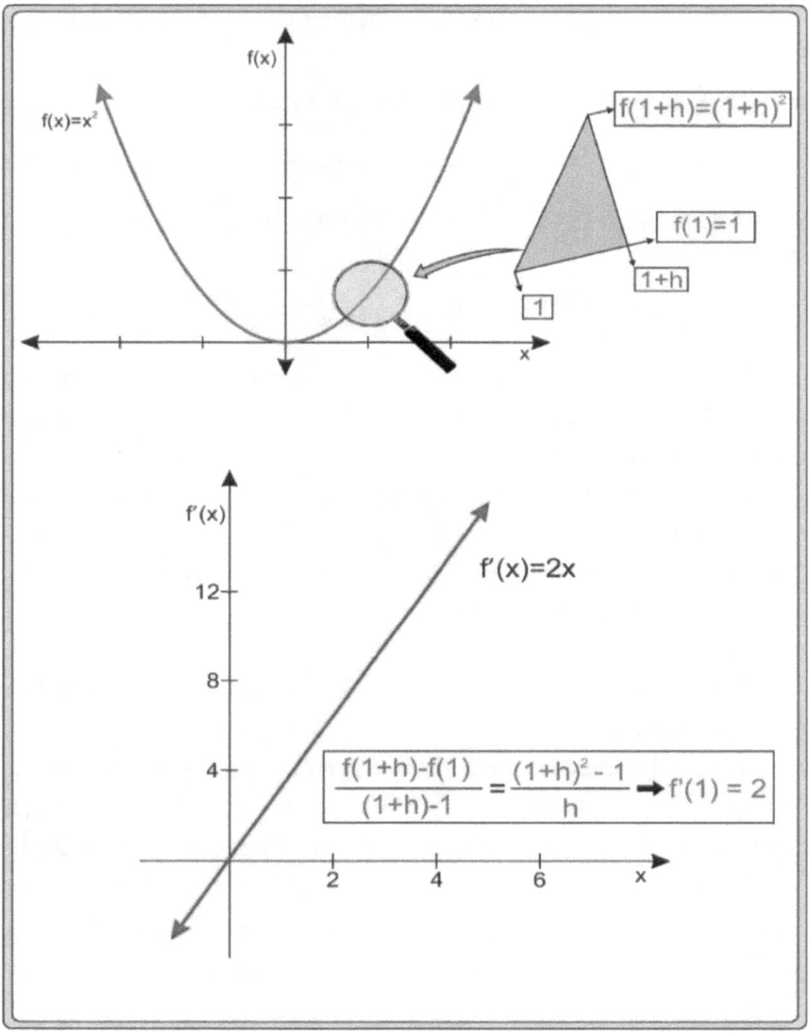

Figure 9.7: An example function explaining how the derivative arises.

$f(x) = x^3$ and $f(x+h) = (x+h)^3$ when you type

$f(x+h) = x^3 + 3x^2h + 3xh^2 + h^3$ using the equations;

$f'(x) = \lim_{h\to 0} \frac{f(x+h)-f(x)}{h}$

$$f'(x) = \lim_{h\to 0} \frac{x^3+3x^2h+3xh^2+h^3-x^3}{h}$$

Now let's get rid of the excess:

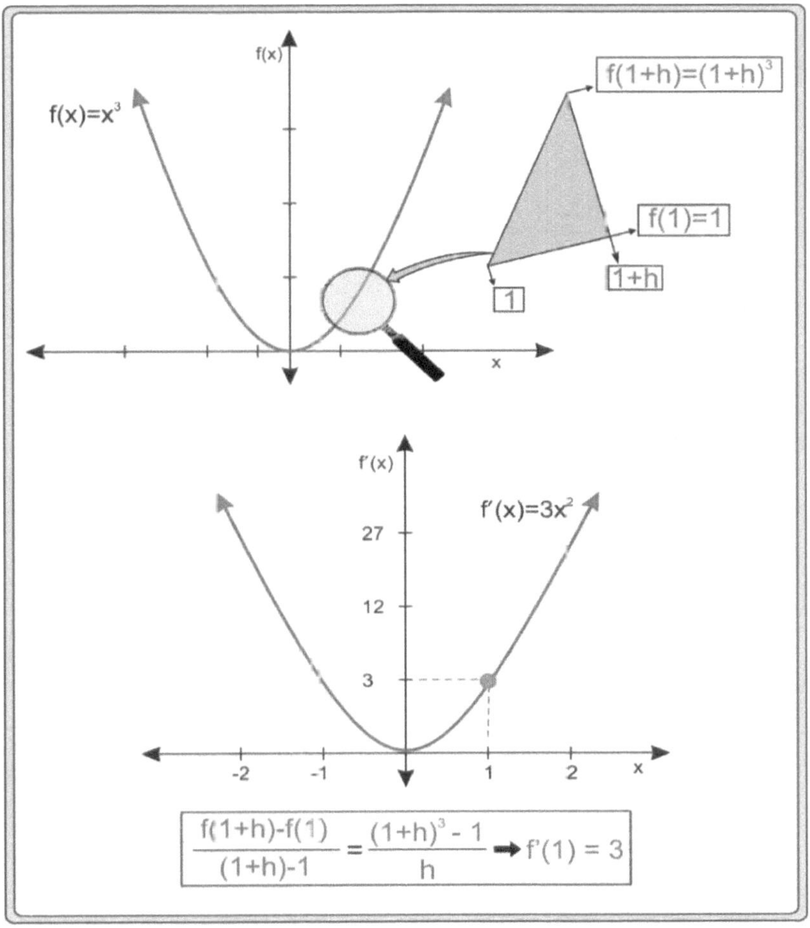

Figure 9.8: Another function explaining how the derivative arises.

$$f'(x) = \lim_{h \to 0} \frac{3x^2h + 3xh^2 + h^3}{h}$$

Also, from here;

$$f'(x) = \lim_{h \to 0} 3x^2 + 3xh + h^2$$ we get.

with the assumption that h=0

$$f'(x) = 3x^2$$ we get.

the values you set for x for 1 and 3 $f'(x) = 3x^2$ function.

You can see 3 for x = 1 and 27 for x = 3, right?

And now, if you like. $sin\ (x)$ Why is the derivative of $cos\ (x)$ Let's find that it is? First, we will use the following equation, which you can find with a simple proof:

$$sin(x+h) = sin(x)\cos(h) + \cos(x)\,sin(h)$$

$$f'(x) = \lim_{h\to 0}\frac{f(x+h)-f(x)}{h} = \lim_{h\to 0}\frac{sin(x+h)-sin(x)}{h}$$

If you edit the bounded expression using the above equation;

$$sin'(x) = \lim_{h\to 0}\frac{sin(x)\,cos(h)+cos(x)sin(h)-sin(x)}{h}$$

You will get it. You remember the limit rules;

For $h \to 0,\ \sin(h) = h\,,\ \cos(h) = 1$ you will find it. Now let's write the limited expression separately.

$$sin'(x) = \lim_{h\to 0} sin(x)\frac{cos(h)}{h} + \lim_{h\to 0}\cos(x)\frac{\sin(h)}{h} - \lim_{h\to 0}\frac{sin(x)}{h}$$

you get.

Hence for h=0 $cos(h) = 1$;

$\lim_{h\to 0}\sin(x)\frac{\cos(h)}{h} - \lim_{h\to 0}\frac{\sin(x)}{h} = 0$ you will find it.

$sin'(x) = \lim_{h\to 0}\cos(x)\frac{\sin(h)}{h}$ for the L'Hospital rule;

$\frac{\sin(h)}{h}$ when you take the derivative of the numerator and denominator separately;

The derivative of $\sin(h)'$ becomes cos(h) and

for h=0 $\cos(0) = 1$ it happens.

The derivative of h , $h' = 1$ equal.

$\frac{\sin(h)}{h} = 1$ is written as Backward $sin'(x) = \cos(x)$ becomes.

As a result, $sin\ (x)$ derivative of the function $cos(x)$ you can see it here.

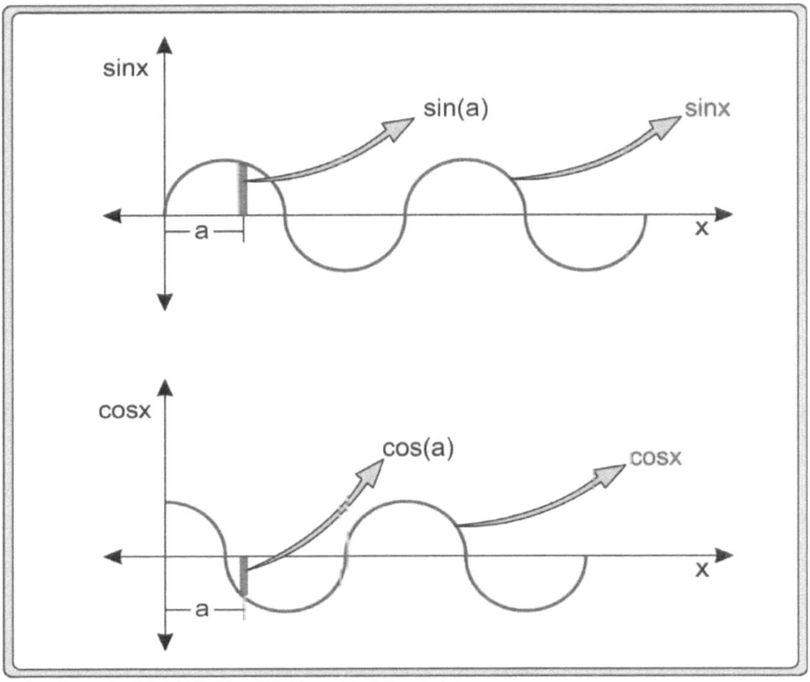

Figure 9.9: Sine and Cosine are true. Do not be deceived by appearances.

Actually here $sin(x)$ The most critical equation we use when taking the

Derivative of $sin(x+h) = sin(x)\cos(h) + \cos(x)\,sin(h)$ is.

Once you prove this, the rest is a piece of cake, isn't it? You can easily deduce how this expression arises from trigonometric transformation expressions.

When we see trigonometric functions, whether derivative or integral, we experience some confusion. The main reason for this is that when we see the expressions sin(x) and cos(x), we see curved shapes rather than straight shapes such as angles, arcs and circles.

I would like to underline once again that the functions sin(x) and cos(x) actually mean the ratio of the diagonal/hypotenuse and the adjacent diagonal/hypotenuse, and that the operation is a division operation, so the functions sin(x) and cos(x) are the ratios of the lines to each other, and that you learn the concept of ratio and proportion to internalize them.

"Now why are we doing division to find the change?" you might be asking yourself. The fact is that, as I mentioned earlier, it is nothing more than knowing the area of a rectangle and the length of one side and then finding the length of the other side. So, to find the speed, we are doing nothing more than dividing the distance by the time. You can see and understand what I mean more clearly when you look at the world of formulas.

The world of functional relationships offers you an environment where you can use both derivatives and integrals in abundance. If you approach them as finding the area of a rectangle, you can see the derivative and integral behind finding the area of a rectangle with known side lengths or finding the path from the rectangle to the edge. I drew your attention to the product of pressure and area when finding the force in the integral, and here I draw your attention to the door to the derivative, that is, when you want to find the pressure, you need to divide the force by the area.

In classical education, it is emphasized that the derivative is the slope. You may remember from the right triangle that the slope is actually the ratio of two lines to each other. While explaining the derivative, I deliberately did not emphasize the slope too much, because the slope is not enough for me to explain the derivative. Of course, we can reinforce the derivative with the definition of slope. With the slope, it is nice to be able to see whether the function is in an increasing or decreasing trend. Just like the acceleration and deceleration of a car, the slope gives you an idea of which way your curve is going. But when you say that derivative means slope, I object to that. When you explain the subject with slope, the information about what the derivative does and where it is used is always hidden. You cannot get out of it by saying slope. If you are going to bring the derivative into life, you need to explain what it is used for.

When I was explaining the integral, I talked about its physical meaning in detail. Of course, you can see the derivative through division operations in formulas. But I want to show you the derivative through functional relationships. It is useful to briefly mention the physical meaning of the derivative here. As you know, physics defines 7 fundamental quantities: mass, length, time, current strength, temperature, light intensity and amount of matter. I have already mentioned that the quantities derived from these are quantities such as force, acceleration, velocity, resistance and energy. You can see the derivative and integral much more easily in these derived quantities.

SI Base Unit	
Fundamental Quantities	SI
Length	m
Mass	kg
Time	s
Electric current	A
Temperature	K
Luminous intensity	Cd
Amount of substance	mol

SI Base Unit	
Derived Quantities	SI
Force	Kgm^{s-2}
Velocity	ms^{-1}
Pressure	$Kgm^{-1}s^{-2}$
Energy	$Kgm^{2}s^{-2}$
Acceleration	ms^{-2}
Resistance	$Kgm^{2}A^{-2}s^{-3}$
Power	$Kgm^{-2}s^{-3}$

Figure 9.10: Units gives you the full picture of derivatives and integrals.

When we were explaining the integral, we talked about the product of quantities. In the derivative, of course, we will focus on the quotient. You may remember the table above when we were explaining the integral before. Now look at this table by focusing on division, not multiplication. Let's see what you can get when you divide the quantities by each other.

If you divide velocity (ms-1) by time - which is the derivative of velocity with respect to time - you get acceleration (ms^{-2}), and if you divide energy (kgm^2s^{-2}) by time, you get power (kgm_2s^{-3}).

A careful look at the table will show you how many of these can increase. Aren't many of the expressions such as speed, path, acceleration, energy, pressure, force already required in the problems? Focus on the units of magnitudes and you will understand better what I mean.

Let's explain the derivative and integral over electrical resistance, the longest magnitude in the table: Look at the unit of magnitude ($kgm^2A^{-2}s^{-3}$); this is our goal. First, we integrate the mass (kg) twice with respect to the length, or once with respect to the area (m^2). Then you take the derivative of these two (kgm^2) twice with respect to electric current (A) and you get the expression (kgm^2A^{-2}). If you take the derivative of these three times with respect to time (s^{-3}), you get the electrical resistance ($kgm^2A^{-2}s^{-3}$). What more can derivative and integral do! Remember. To find electrical resistance, the most difficult quantity in the table, all I do is multiplication and division.

I would like to draw your attention to another issue here: derivative and integral can actually be the same thing. That is, the derivative of one function can be the integral of another function. For example, speed is the derivative of the path, while acceleration is the integral. I would like to emphasize here that the derivative and the integral are actually nothing more than the tools you use to climb or descend a ladder. Just as you reach a goal by climbing or descending a ladder. Just like that, you can transform a function into the function you want by division and multiplication. We just call this process differentiation or integration.

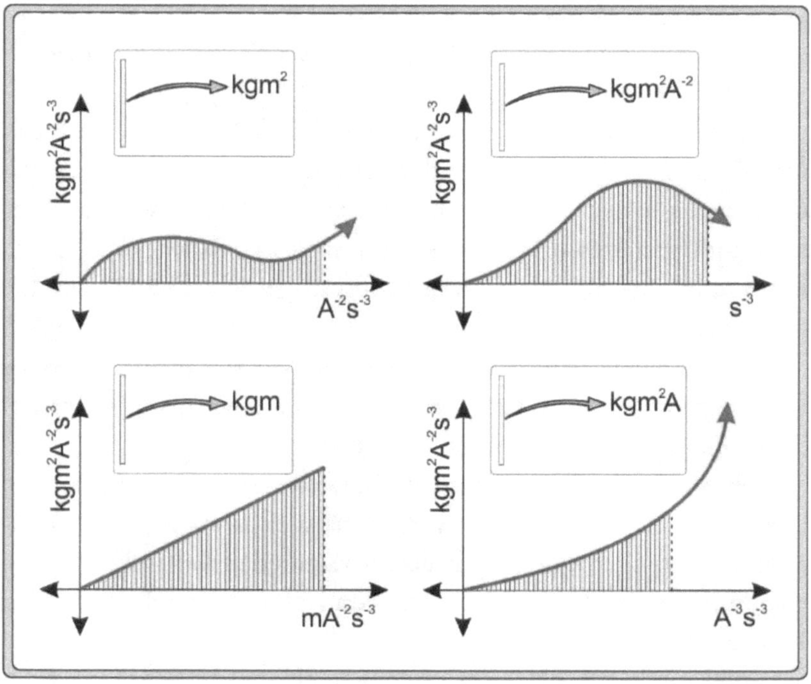

Figure 9.11: Graphs where the derivative is clearly seen through a formula.

At the beginning of the chapter, I said that it is better to teach integration first and then derivatives. If I see a graph of a curve in a coordinate system, I want to calculate its area first. For example, if I see a velocity-time graph, I want to calculate the path immediately, and then I think of acceleration. That's why it seems more accurate to me to explain the integral first. That's why I explained it first.

The most important reason for explaining derivative and integral concepts through equations of motion is that position, velocity and acceleration are concepts that travel through your life and can be understood by everyone. It is thought that it is easier to understand derivatives and integrals through their graphs. I think you can use the concepts of derivative and integral to understand many different phenomena without looking at the subject only to analyze the motion of a car, and I think you will better understand what I mean when you look at the table where I give some basic and derived physical quantities.

Unfortunately, the number of events that we observe in our lives is so small that we perceive the subject of motion in physics classes as solving only the problem of speed, and we miss the fact that this subject is actually chosen only as an example to analyze a picture whose functional relationship is determined and graphed. If you can see the real meaning of position, velocity and acceleration, you can easily establish and solve the transformations between other functions.

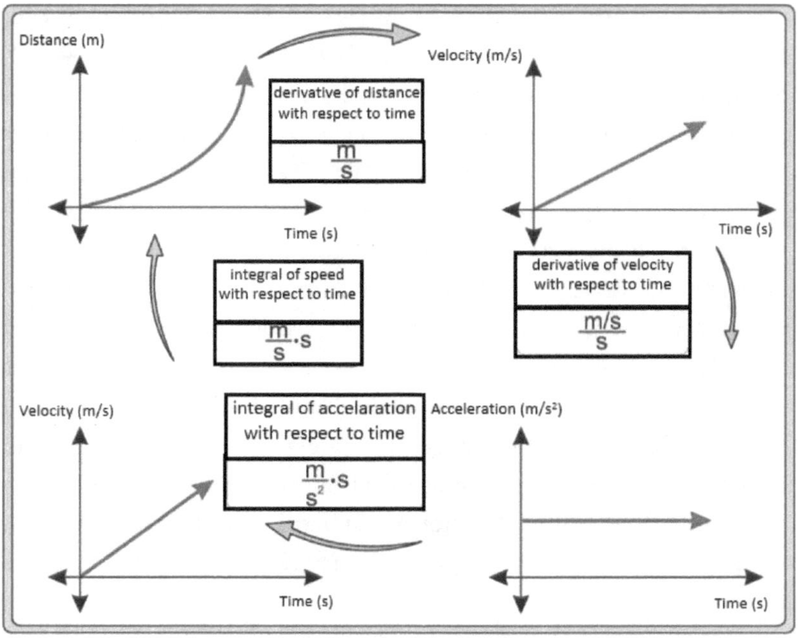

Figure 9.12: Explanation of derivative through path-velocity-acceleration graphs.

When you consider the time-dependent change of position, velocity and acceleration, as well as the fact that objects are three-dimensional, that is, when you combine the concepts of edge, area and volume, you can see that motion has an equivalent in almost every part of our lives. In fact, motion is the name of change and it is a physical event that touches our lives the most. The most important independent variable that makes movement meaningful is time. We can talk about motion wherever there is change with respect to time. Therefore, motion is very valuable for understanding and interpreting derivatives and integrals.

We can say that there is almost nothing around us that remains constant. The deterioration times of materials in cold weather and materials in hot weather are very different. Cars that stay in hot and humid regions wear out much more quickly than others. All of this is the work of a movement. Change starts with movement. If you can analyze this effect properly, you can overcome other problems very easily.

The time a yogurt lasts in the refrigerator will be different from the time it lasts in hot weather. If you want to draw meaningful conclusions from these changes, the derivative will be your biggest assistant. Just as you constantly check your speed while driving a car, knowing, controlling and interpreting the speed of change is one of the most important principles of analytical thinking. There are so many facts that change will tell us that the volume of this book is not enough to tell them all. If you internalize this physical phenomenon, then it will not be difficult for you to comprehend and interpret the others.

As you know, we learn derivative and integral with infinitesimal concepts called dx and dy. If they are infinitesimal concepts, then how do we perceive their differences? Everything infinitesimal is similar. Remember the limit, we process everything by rounding. So the difference between everything that gets smaller disappears. If this is the case, then we think that there is no difference between dx and dy, that is, they are the same.

And this is where the problem comes in. You have to see that they actually represent infinitesimally different infinitesimals, just like atoms. Just like the atoms of each element are different... You can't just add and subtract, multiply and divide them. In the same way, dx and dy are different from each other. You can't just add, subtract, multiply and divide them. This is how you need to learn these concepts while perceiving the spirit of mathematics. Mathematics is built on a philosophy so broad that it can absorb all the details of physics. If you read and interpret them from this point of view, then you will better understand what mathematics means.

As a result, you may define thousands of relationships between measurable quantities, as exemplified in the table where I have given some fundamental and derived physical quantities, and you may want to create and solve sets of questions such as "What has changed relative to each other, how has it changed, what are the effects?". Here, the derivative is the most important mathematical function that helps you to analyze them. That's why I'm glad there is a derivative!

Matrix

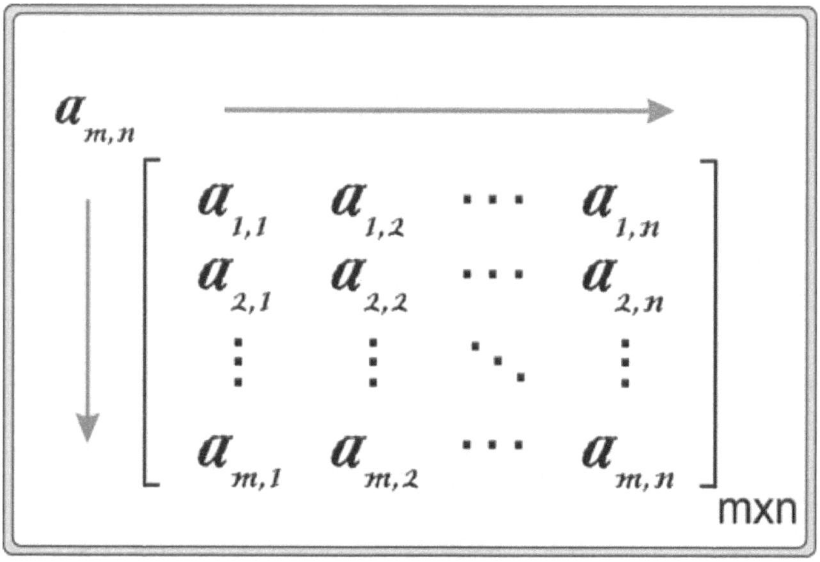

Why Learn Matrix?

I met the matrix in my last year of high school. I remember that our teacher did not explain the matrix very quickly and as if it was a very important subject in order to catch up with the curriculum. Of course, we learned the rules of the matrix like other subjects. We learned how to add, subtract, multiply, take the inverse and find the determinant of a matrix, but like other subjects of mathematics, we graduated without getting an answer to the question "Why did we learn these things?".

Isn't it strange to be unable to answer questions like "Why do we take the determinant of a matrix or what is the use of the determinant we take?" Think about it, you spend years learning and at the end, when you are asked why you learned this, you cannot answer properly. Don't expect PISA results to change until this strange situation is fixed!

I don't want to be modest about solving problems in math classes. But it was only years later that I realized that when it came to basic questions such as, "Where did this come from and what is the equivalent of this in life?" When I talk about these issues with the people around me, I can easily say that unfortunately, we are all in more or less the same situation. The fact that we do not have the answers to these questions is the clearest indication that mathematics does not touch our lives.

When you look at engineering problems, we can say that it has emerged as a need to see multiple results of multiple effects entering a system; in short, to be able to process and analyze multiple functional relationships and data at the same time. Because life is not lived with one unknown, but with multiple unknown relationships.

When you look deeply into this life, you can easily see that we live a life that can be defined by multiple functional relationships. It all depends on your perspective and the deeper you look into this life, the more functional relationships you have. It is very difficult to overcome them with classical mathematical operations. The matrix provides you with very effective solutions for analyzing these multiple functional relationships. I will try to explain why and how in this part of the book.

I have been talking about functions since the beginning of the book. I have talked about functions everywhere. When you discover the magic world of the matrix, you realize once again how important functions are. As I keep underlining, you enter the virtual world, the mathematical world, from the real world with functions.

Functional relationships are defined by two basic quantities called scalar and vector. The analysis and synthesis you do with these quantities using the four operations is the essence of mathematics. All mathematical concepts are built on these quantities.

We first start to learn functions with polynomials. However, functions defined by polynomials are not enough to solve real problems. So we start to use logarithms and trigonometry and functions that allow us to define more complex relationships. We try to solve problems by adding, subtracting, multiplying and dividing them. As these operations become more complex, other important tools of mathematics such as limits, integrals and derivatives come to the rescue.

When it comes to matrices, we are actually looking for solutions in a world of problems with a few small functional relationships. In the world of two- and three-dimensional problems, it is not very difficult to find solutions without using a matrix. Here, the matrix comes to the forefront as a mathematical tool that allows multidimensional functions to be analyzed using four operations, where many more relationships are defined on top of all these teachings.

Now let me explain what I mean with a simple example:
f_1 (x,y,z)= 3x+5y+8z
f_2 (x,y,z)= 5x+4y+7z
Let f_3 (x,y,z)= 4x+6y+7z be functions.

Each of the above functions can have a meaning in itself. Since I have already explained the functions, I will not explain again what the numbers and symbols mean. You can think of these functions as part of a common problem. For example, when we sit, we can yawn and sway from side to side. Remember that at the same time you are breathing in and out and our digestive system continues to work! We live a life under the joint influence of all these relationships. The matrix is a mathematical concept created to make such multi-input and multi-output relationships understandable. Thanks to the smooth structure and ease of representation of the matrix, you will not get lost in numbers and figures when analyzing problems.

$$\begin{bmatrix} 3 & 5 & 8 \\ 5 & 4 & 7 \\ 4 & 6 & 7 \end{bmatrix} * \begin{bmatrix} x \\ y \\ z \end{bmatrix} = \begin{bmatrix} f_1(x,y,z) \\ f_2(x,y,z) \\ f_3(x,y,z) \end{bmatrix}$$

Thanks to the matrix, you can solve problems like the one above for the joint effect of three different functions. These problems can be the deformation of a material subjected to heat and pressure under different loads, or a problem analyzing the vibration on the wing of an airplane subjected to loads at different altitudes. Thanks to the matrix, you have the opportunity to simultaneously examine and analyze the effect of not only pressure on a geometry, but also the effect of pressure, temperature and fatigue in the material, or the effect of another condition that occurs when it rains. Depending on your approach to the problem, the size of your matrix increases as you multiply these relationships.

Take for example the density of air, which varies with altitude. You can write the density of air as a function of altitude. Temperature can also be a parameter that affects the density of the air. You live in the same place; the pressures may be different in summer and winter. Then you need to consider the effect of temperature on the pressure function. Remember the nested functions! This time you can write temperature as a variable of season, month, day, time. It is the basis of engineering to find out what the pressure is at what altitude and at what time and to use it in a problem where both altitude and temperature are considered as variables. You will understand better what I mean when you see that the number of your functions will reach millions when you calculate the effect of pressure on the surface, for example on an airplane wing, by dividing the structure into small parts.

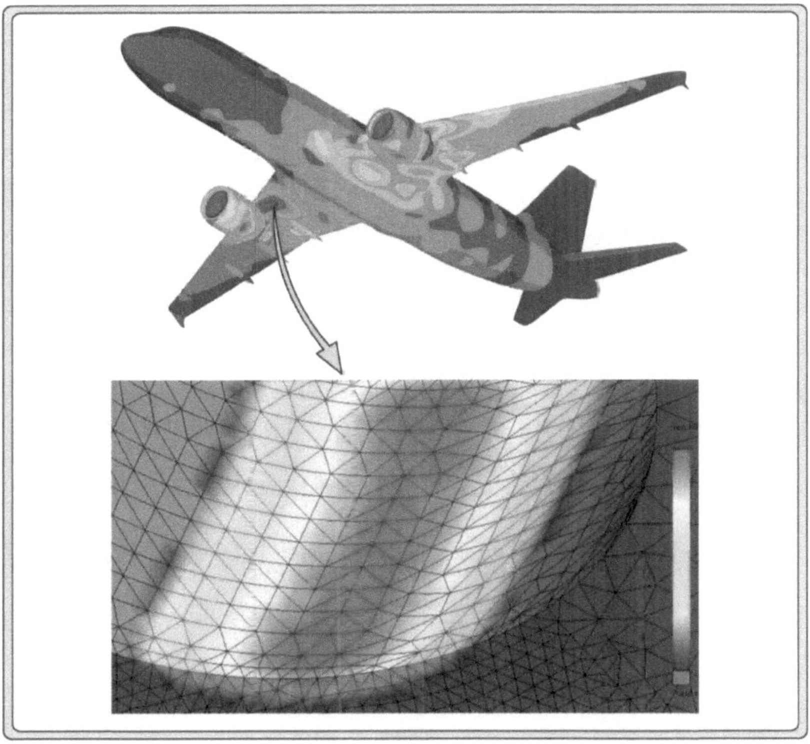

Figure 10.1: Where thousands of functional relationships meet the matrix: Finite elements.

In a nutshell, the matrix allows you to see in detail that the effect of hundreds, thousands, even millions of variables on a system can also have hundreds, thousands, millions of consequences. The matrix is a mathematical representation that allows you to make an evaluation between them and make decisions. In fact, as it can be understood from here, the matrix is a laboratory where multiple functions are analyzed.

In the engineering world, there is a design and analysis program called MATLAB that is widely used by many engineers. MATLAB stands for MATrix Laboratory. As the name implies, it is a program designed for the use of matrices. I think the name of this program is enough to explain the importance of the subject. Because there is no such program as an integral or derivative laboratory, but there is for matrices. The reason for this is that in real problems there are often more than one functional relationship and these can be represented as a table of numbers thanks to the matrix. In this way, the confusion is eliminated. Finding and extracting meaningful expressions from this table of numbers, which we call a matrix, is the best method until a better mathematical representation is found.

In the days before powerful calculators such as computers, of course, it was not very easy to analyze thousands of functional relationships by hand calculation. Therefore, in the pre-computer era, a lot of assumptions were made and functions were simplified to solve problems. Therefore, the results obtained here were often unrealistic.

Figure 10.2: See the difference between two pictures.

It is not possible to make a realistic analysis with too many rounded expressions. We can liken this to the resolution of a camera. When the resolution of the photo you take decreases, your photo looks very little like the real thing. The quality of the image improves as the resolution increases. Just like this, analyzing a structure by dividing it into smaller parts has always yielded more realistic results. Otherwise, of course, you can define your problem with a simple function with one unknown, but this does not show you the real picture. Thanks to the matrix, you have the opportunity to take and analyze much clearer photos, considering the effect of small parts.

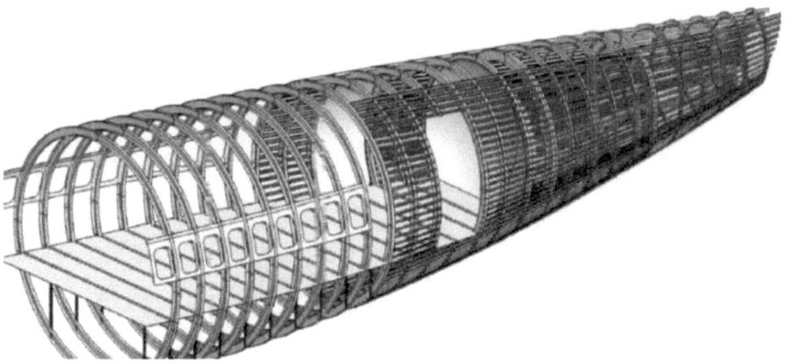

$$K=\begin{bmatrix} 5800 & 0 & 180 & 94 & \cdots & 90 \\ 3200 & 185 & 45 & 0 & \cdots & 80 \\ \vdots & \vdots & \vdots & \vdots & \ddots & \vdots \\ 1800 & 1050 & 120 & 38 & \cdots & 180 \end{bmatrix}$$

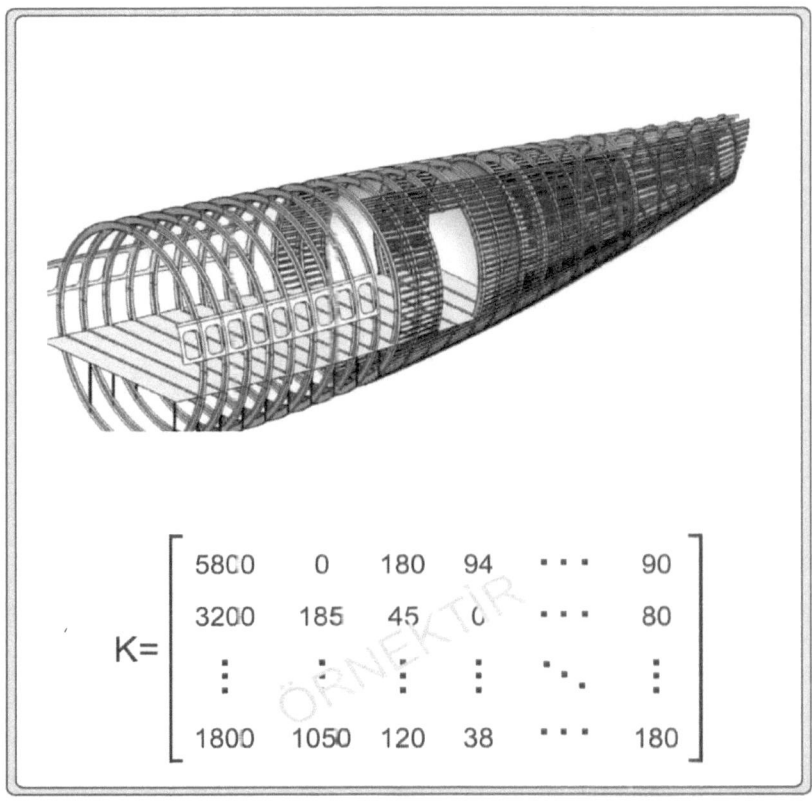

Figure 10.3: We get out of the lattice structures in the airplane thanks to the matrix.

When I took a finite element course at the university, I better understood why there was a need to operate with a matrix. Seeing that the load on the structure is divided into millions of parts with finite elements and that millions of different effects of this are processed at the same time gives the closest results to reality, and that this can only be solved with a matrix, made me better understand the value of the matrix.

With the finite element method, which is widely used especially in aircraft, creating a virtual lattice structure with a network of small triangles on a structure and then analyzing the behavior of the structure under load in detail has enabled the removal of many more unnecessary parts on the aircraft. Thanks to this method, airplanes that carry more load but are lighter have been built. Of course, all this was made possible by the analysis using matrices.

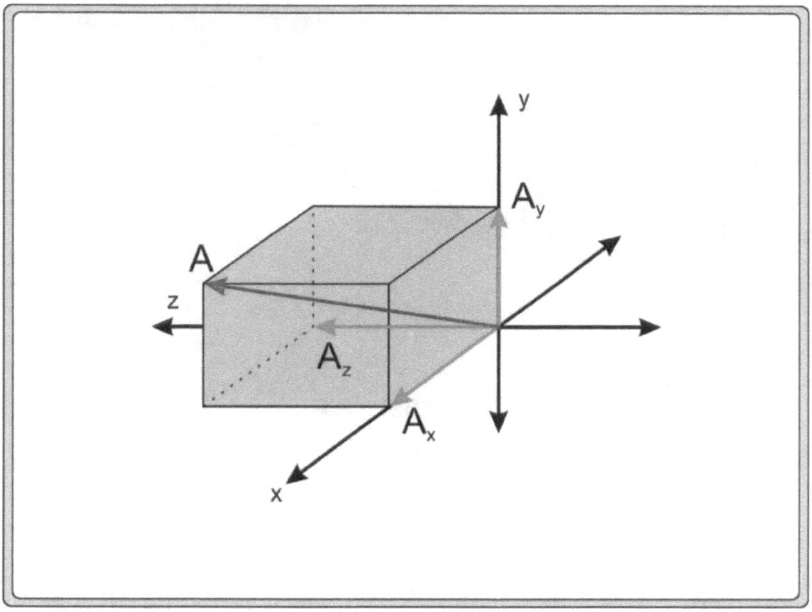

Figure 10.4: Inside matrices, vectors are hidden.

Now I can hear you saying, "Did you find the real meaning of the matrix in functions?" You will be surprised, but my answer to that is no, because I found it in vectors. Vectors, you know the expressions with the arrow on the symbols... This arrow used to annoy me, I used to say, "What is it doing in mathematics?" I didn't even want to understand it. It was only much later that I discovered that the arrow was just a symbol and that it was there to pay attention when doing four operations.

I was even more surprised when I discovered that vector quantities, that is, quantities where the direction also enters the problem, constitute a large part of the world of real problems encountered in life. Unfortunately, for years I had thought that vectors were limited to force vectors in physics classes. But vectors were so many that they were the most fundamental source of the rules of the matrix, and unfortunately, I had not seen this for years.

If we compare the problems to the world we live in, we should compare the problems to be solved with scalar operations to continents and the problems to be solved with vectors to oceans. Because with vectors, the concept of direction also comes into play and the mathematical models of the problems are constructed to make them much more similar to real life.

Examine some of the vector and scalar quantities given below. We write tens, even hundreds of formulas with scalar products of seven fundamental quantities. Now imagine their vector magnitudes. The world of formulas derived with vector products has opened the way for us to establish a much wider network of relationships.

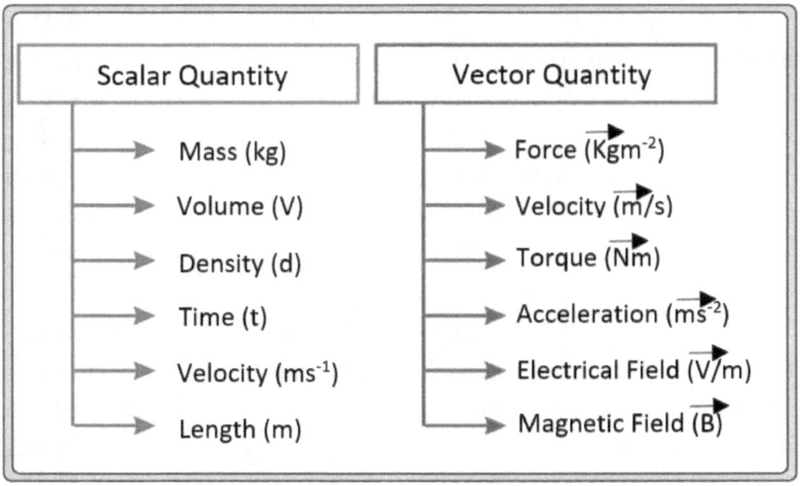

Figure 10.5: Example scalar and vector quantities.

Although the matrix originated to see the effect of more than one function, it has become more understandable in vectors. We all remember force vectors, especially in physics lessons. These vectors tell us how important direction is. But don't think that vectors are limited to these! To perceive vectors, remember the magnetic field, acceleration, velocity and the displacement vector, and take another look at the physical formulas and you will better understand what I mean.

In fact, even time is a vector magnitude. We are not so much interested in its vectorial aspect, we are more interested in its scalar aspect. Maybe there is a vectorial aspect to most quantities that have been defined so far, even though we can't see it. If we can find and reveal this aspect, we will realize that perhaps many more problems than the ones described so far are waiting for us.

Look carefully at the table on the previous page; I can give you so many formulas and equations involving force, velocity and acceleration that your mind will be boggled. There are so many problems involving force, velocity, acceleration, and of course electric and magnetic fields that touch our lives, and you will see that there are always vector quantities in them. Therefore, in real life, vector problems are much more quantitatively than scalar problems. When you see that the areas and volumes of geometric shapes formed in the coordinate system by the introduction of vectors into problems are often the solution to a problem, you will better understand what I mean. I explain that the area or volume you actually find gives you the total force, moment or distance traveled.

You can consider scalar problems, that is, problems where the direction does not matter, as simple problems that have entered our lives since primary school. If 1 apple is 5 liras, how many liras are 4 apples, or how many kilometers a vehicle that travels 60 km in 1 hour travels in 2 hours, we can show these quantities on a number line and find the solution on the number line.

Vectors are different from scalar quantities and are actually where algebra and geometry meet. Because we cannot perceive the concept of direction without using geometry. Vectors not only carry your problems into three-dimensional space, but also open the door to infinite dimensional space thanks to the matrix. Take a look at the formulas derived so far and you will see three basic operations: the vector product of a vector and a vector, the scalar product of a vector and a vector, which is also the scalar product of a scalar and a scalar, and the product of a scalar and a vector. And that's all there is to it! These multiplication operations tell us how much this world can expand thanks to vectors. Only one of the three products is a scalar product. That's why I liken the problems to be solved with scalar operations to continents, and the problems with vectors to oceans.

In various chapters of this book I keep reminding you of the basic formulas in physical problems because it all started with analyzing them. There you see the most basic mathematical expressions: addition, subtraction, division and multiplication, the four operations. You also notice in the formulas both scalar and vector quantities. The difference, sum, product and quotient of vectors are the most common operations we encounter in problem solving. Therefore, when we understand what vector expressions mean, we are one step closer to the spirit and rules of the matrix. To understand this topic, let's first take a look at the difference between vector product and scalar product.

First, let's take a look at what scalar multiplication is. This is the name of the multiplication operation that can be represented on the number line. From primary school onwards, we try to solve relationships in the world of numbers using the number line. Analyzing scalar quantities on a straight line makes our work easier. With scalar multiplication, you get meaningful expressions on a line, while with vector multiplication you get meaningful expressions that open to the plane and of course to 3D space.

Take a look at the scalar product of vectors, and you will see that the second vector is multiplied by the magnitude obtained by taking the cosine, the projection of the first vector with respect to the second vector. Why cosine? Because by multiplying by cosine you bring both vectors in the same direction. You already know that scalar multiplication is a simple multiplication operation that we use to multiply expressions in the same direction, and that it is explained on the number line.

Now go back to your elementary school classrooms and remember again why you add, subtract, multiply and divide on the number line. Then you will once again witness that all mathematical concepts, from derivatives to integrals, from vectors to matrices, have emerged in order to analyze all the formulas from $F=ma$ to $e=mc^2$ in order to make sense of them.

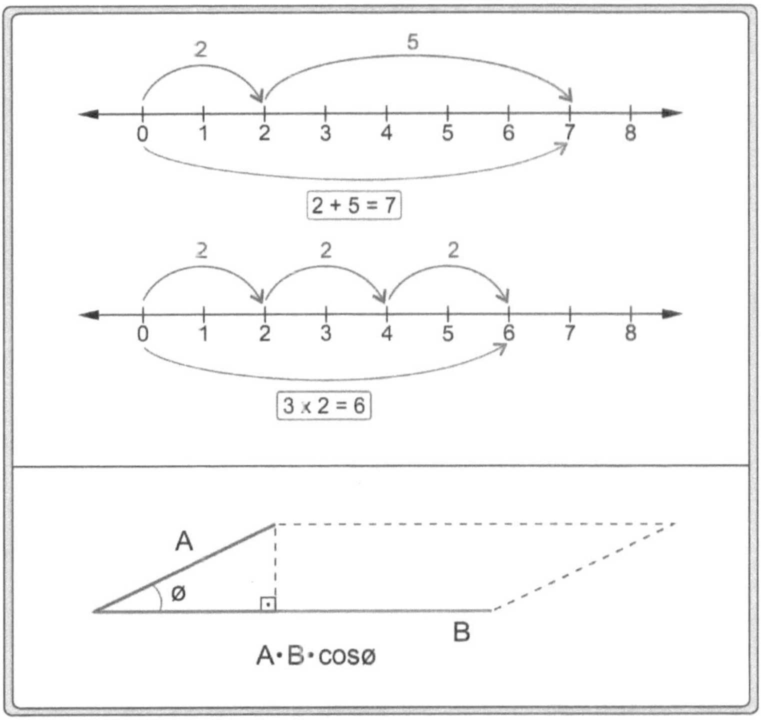

Figure 10.6: Four operations on a scalar make sense on the number line.

In fact, cos(0) seems to be hidden inside scalar quantities like a hidden subject. Since cos(0) corresponds to 1, it is treated as an ineffective element in scalar multiplication. Even though it is hidden, there is a cos(0) in all scalar quantities.

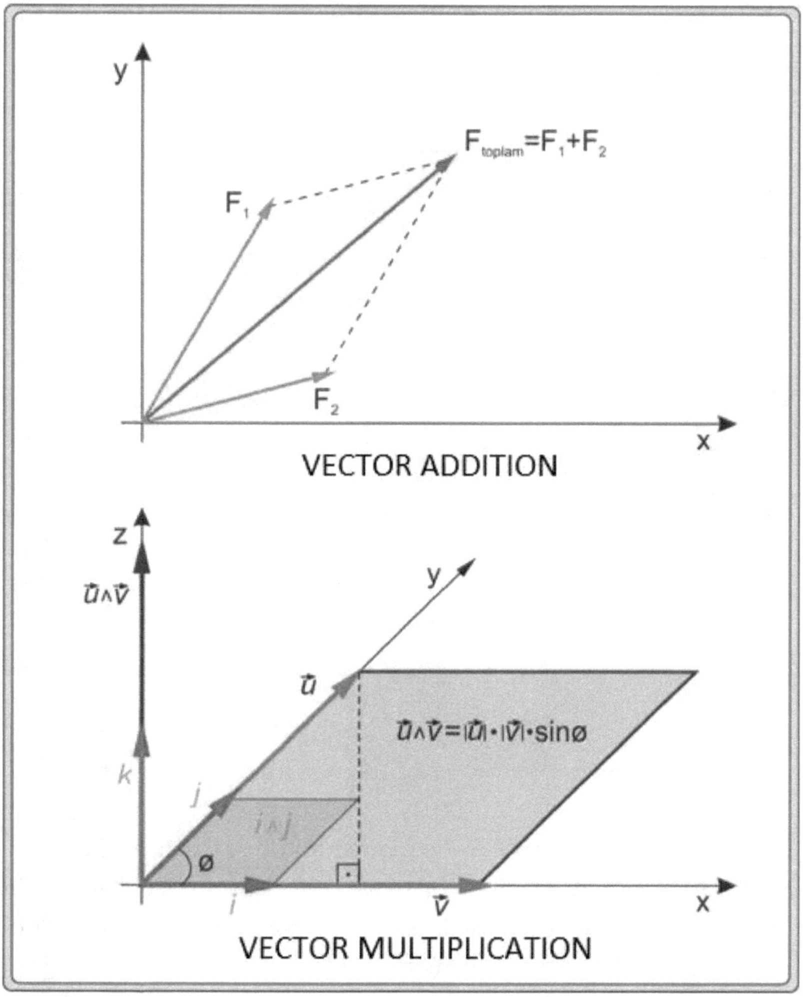

Figure 10.7: Vector operations are interpreted in three-dimensional space.

You can start to understand vector multiplication by finding the area of a rectangle. While the scalar product finds its meaning on a number line, i.e. the result of the product can be represented on that number line, the vector product can be visualized by the area of a surface. If the vectors are perpendicular to each other, the product gives the area of the rectangle in between. If they are not perpendicular, i.e. there is an angle other than a right angle, then this shape will look like a parallelogram and you will find its area.

Rectangles come up all the time when learning differentiation and integration. When I was learning integration, I explained that the area of a rectangle can correspond to the perpendicular line of another function. The basic logic of this is hidden in the statement "The area of one function can be the edge of another function". Whether you differentiate or integrate, what you are doing is nothing more than transforming functions into each other. The spirit of Linear Algebra also appears here. Therefore, when you multiply vectors by a vector, the area you find is also represented by another vector magnitude.

If the vectors are not perpendicular to each other, i.e. if there is an angular relationship between them other than a right angle, of course the resulting shape will resemble a parallelogram. You get the area of the parallelogram from the product of the first vector and the sine of the second vector. Now you may ask, "Why take the sine?" The answer is that you multiply the base by the height to get the area of the parallelogram. That is why the sine comes into play in vector multiplication. By taking the sine, you get perpendicularity and from there you can easily get the area.

Of course, if you multiply this area by a perpendicular line segment, you get the volume. Think of three-dimensional geometric shapes like cubes and prisms. Concepts such as area and volume, which are used extensively in functional analysis, become even more meaningful with vector multiplication. This product, of course, continues to expand. As the dimension increases, geometry, the coordinate system, can no longer help you visually, but the matrix gives you the opportunity to pick up where you left off. This is the spirit of vector multiplication. The foundations of the matrix are laid here.

To understand this better, review the formulas used in physics classes. You will see the three basic multiplication operations I mentioned above there in abundance. Of course, you need to deduce their physical meaning from the expressions in the formula. In other words, if you know what the multiplied elements in a formula mean, you will understand what the expressions there turn into. Just remember that you are in the mathematical world when you are analyzing or interpreting a problem, whether it is on a number line or whether the multiplication turns into an area or volume. If you pay attention to your definitions when transferring a physical problem to the mathematical world, you will not be separated from reality. But if you stay only in the mathematical world, you cannot see the whole reality. So you have to go in and out of the physical world once in a while.

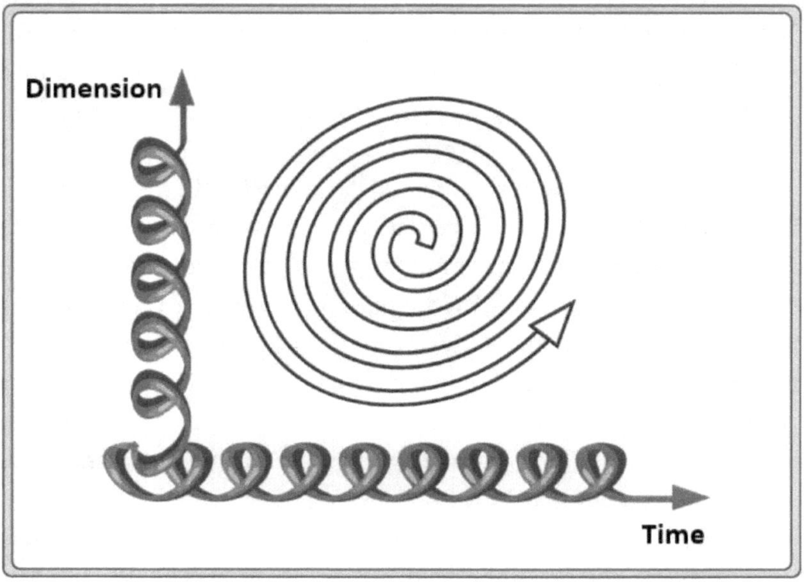

Figure 10.8: Time and dimension are non-linear. We may need different coordinate systems to understand them better.

Vectorial or scalar expressions are related to the definition of your problem. For example, speed is a vector quantity while time, which is often used in problems, is a scalar quantity. The product of these, i.e. the product of a scalar and a vector quantity, of course gives a vector quantity. We take time into problems as if it flows on a number line, but maybe time is not linear but spiral. If you take time as a vector quantity, you see that the product of two vector quantities is also a vector quantity. For now, we don't use the vectorial aspect of time to solve problems. Perhaps in the future we may come across problems where we will also do this kind of calculation.

Also remember that each vector magnitude has a scalar magnitude! So, speed is a vector magnitude, but you can also solve problems with the scalar side of this magnitude. Then, to avoid confusion, you may need to call it something else, usually speed. If it doesn't matter where you are going, multiplying the scalar magnitude of the speed by the elapsed time will tell you the distance traveled, which is a scalar expression. If the direction you are going is important, simply multiply the vector magnitude of the velocity by the scalar magnitude of the time. From this multiplication you get the displacement vector. In some problems, the direction is not important and only the distance traveled is required. In some problems it may also be important where you are going, which is often more important in real problems, in which case you will be asked to find the displacement vector.

When we give an example of a physical event, we immediately think of the triad of acceleration, speed and path. Don't always perceive this triad in a physical event like traveling by car. When you look carefully at life, you realize that this triad appears in every physical event where there is change. The critical word here is change. Remember. If change has speed, it has acceleration, it has a path.

The vector symbol in the formulas is very confusing. In fact, the vector symbol, as I have explained before, is only used to show that you need to be careful when doing four operations. In other words, this arrow used in formulas means "Be careful, don't do scalar operations in the classical way!" when multiplying, dividing, subtracting and adding. It has no other meaning. Just remember this when you see the arrow.

To understand the vector product, let us define two simple vectors. First vector on the coordinate system $\vec{u} = u_1\vec{i} + u_2\vec{j} + u_3\vec{k}$, the second vector $\vec{v} = v_1\vec{i} + v_2\vec{j} + v_3\vec{k}$ as

In order not to confuse vector multiplication with other multiplication operations, we define the multiplication operation with the symbol $\wedge$. Let's pay attention to the multiplication operation here. Now let's multiply them: $\vec{u} \wedge \vec{v}$ You will see that this is done by taking advantage of the distributive property of multiplication over addition that we learned in primary school.

Let us define two three-dimensional vectors as follows as an example;

$$\vec{u} = (u_1, u_2, u_3) \quad \vec{v} = (v_1, v_2, v_3)$$

Scalar products are written as follows:

$$\vec{u}\,.\,\vec{v} = (u_1\vec{\imath}, u_2\vec{\jmath}, u_3\vec{k})\,.\,(v_1\vec{\imath}, v_2\vec{\jmath}, v_3\vec{k}) =$$
$$u_1v_1\overrightarrow{\imath.\imath} + u_1v_2\overrightarrow{\imath.\jmath} + u_1v_3\overrightarrow{\imath.k} + u_2v_1\overrightarrow{\jmath.\imath} + u_2v_2\overrightarrow{\jmath.\jmath}$$
$$+u_2v_3\overrightarrow{\jmath.k} + u_3v_1\overrightarrow{k.\imath} + u_3v_2\overrightarrow{k.\jmath} + u_3v_3\overrightarrow{k.k}$$

In scalar multiplication, quantities in the same direction are multiplied, the product of quantities on other axes is equal to zero. Therefore, when the expression is simplified, it becomes as follows.

$$\vec{u}\,.\,\vec{v} = u_1v_1\overrightarrow{\imath.\imath} + u_2v_2\overrightarrow{\jmath.\jmath} + u_3v_3\overrightarrow{k.k}$$

We can write the vector product as follows:

$$\vec{u} \wedge \vec{v} = u_1v_1\overrightarrow{\imath.\imath} + u_1v_2\overrightarrow{\imath.\jmath} + u_1v_3\overrightarrow{\imath.k} + u_2v_1\overrightarrow{\jmath.\imath} + u_2v_2\overrightarrow{\jmath.\jmath}$$
$$+u_2v_3\overrightarrow{\jmath.k} + u_3v_1\overrightarrow{k.\imath} + u_3v_2\overrightarrow{k.\jmath} + u_3v_3\overrightarrow{k.k}$$

In vector multiplication, the product of quantities in the same direction is zero. If the products of quantities on other axes are arranged according to the right-hand rule;

$$(u_2v_3 - u_3v_2)\vec{\imath} + (u_3v_1 - u_1v_3)\vec{\jmath} + (u_1v_2 - u_2v_1)\vec{k}$$
$$= (u_2v_3 - u_3v_2\,, u_3v_1 - u_1v_3\,, u_1v_2 - u_2v_1)$$

equality is obtained.

I think it goes without saying that in the vector multiplication of two vectors facing the same direction, the result should be zero because the angle between them is zero. So, when you multiply i by i, j by j, k by k, you will see that the result will be zero. It is also worth remembering that when you multiply i by j, you get k; when you multiply j by k, you get i; when you multiply i by k, you get j; and when the order of multiplication changes, you get -i, -j and -k, and this is found by the right-hand rule. Notice that scalar multiplication and vector multiplication complement each other.

Order is important in vector multiplication. If the first vector and the second vector are interchanged, the result of the multiplication is an expression in the opposite direction, even though the magnitude is the same. The multiplication rule of the matrix is based on this, and therefore the multiplication process in the matrix is organized in order. In other words, the spirit of vector multiplication is behind the fact that multiplication with the first matrix and the second matrix swapping places gives a different result.

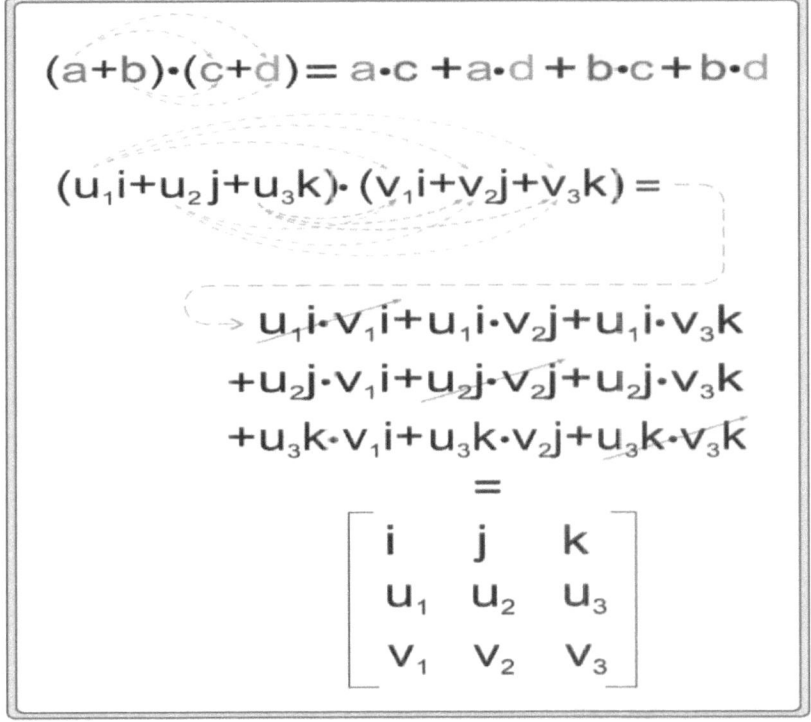

$$(a+b)\cdot(c+d) = a\cdot c + a\cdot d + b\cdot c + b\cdot d$$

$$(u_1 i + u_2 j + u_3 k)\cdot(v_1 i + v_2 j + v_3 k) =$$

$$\begin{aligned} &u_1 i\cdot v_1 i + u_1 i\cdot v_2 j + u_1 i\cdot v_3 k \\ &+ u_2 j\cdot v_1 i + u_2 j\cdot v_2 j + u_2 j\cdot v_3 k \\ &+ u_3 k\cdot v_1 i + u_3 k\cdot v_2 j + u_3 k\cdot v_3 k \end{aligned}$$

$$= \begin{bmatrix} i & j & k \\ u_1 & u_2 & u_3 \\ v_1 & v_2 & v_3 \end{bmatrix}$$

Figure 10.9: In matrix multiplication, the multiplication has the property t hat the product is distributed over the sum.

If you organize vector multiplication, you can see that there is a door to matrix multiplication and at the same time to the determinant, for this, when you decompose the two vectors and multiply them in turn, you can see that matrix multiplication becomes meaningful with the property that multiplication is distributed over addition. We are coming to the end of the book, but let us emphasize once again that mathematics does not go beyond the limits of the four operations.

I am trying to explain the matrix in terms of vector quantities. A matrix is of course a table of numbers and a representation that finds its meaning in a coordinate system. The matrix is not interested in whether the problem is vectorial or scalar. As you can see from vector multiplication, the matrix takes most of its rules from vector multiplication, but this does not mean that the matrix can only be used to solve vector problems. Because all the rules of the matrix make sense through the coordinate system.

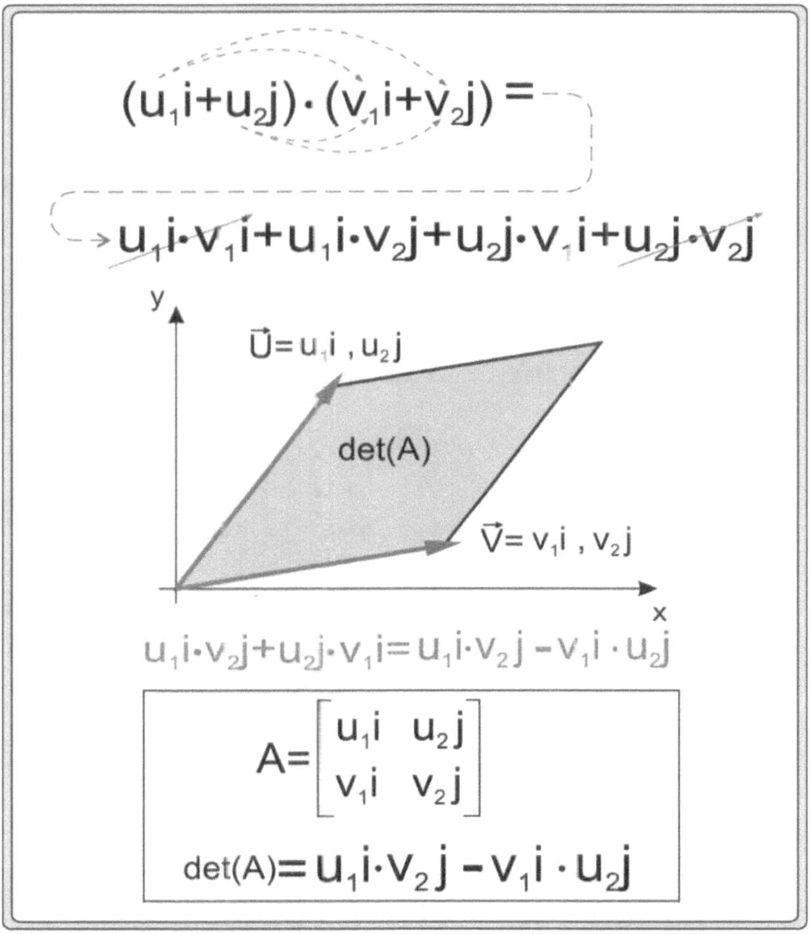

Figure 10.10: Illustration of determinant.

There is something you should pay attention to here: The expressions that flesh out the line equations in the coordinate system are the coefficients in front of the axes. Lines, as you know, consist of lines going to infinity. In a line equation like ax +by=c, you are interested in what the coefficients a, b and c are. If I ask you to write the equation of a line, all you have to do is to find these coefficients.

As you know, number lines have a direction that flows to infinity. When you want to obtain meaningful expressions in the coordinate system using the coefficients a and b, the two most important variables of a line equation, you convert the coefficients a and b into points in the coordinate system. Using these coefficients, which define number lines, to operate with points that turn into finite expressions in the coordinate system is a great beauty that mathematics offers you. The concept of infinity, which describes the number line, becomes dependent and meaningful through these finite expressions. It is through all these dependency relations that you get meaningful expressions with the matrix.

What I have tried to explain so far is that the coefficients describing lines going to infinity, whether vectorial or scalar, are converted into finite expressions with a matrix to obtain meaningful results. The matrix elements can sometimes be the coefficients of a vector and sometimes the coefficients of a line equation describing a scalar problem. Both of these mean the same thing in the coordinate system. What is important here is that these coefficients or point coordinates are actually dependent expressions that can be easily transformed into each other. Being able to use the coefficients of a line equation to obtain the coordinates of a point and establish a meaningful relationship between them is an important feature that analytic geometry offers you.

Let's explain the difference between scalar and vector multiplication here. If you multiply vectors scalarly, you can see that quantities on the same axis can be multiplied, while quantities on different sets of axes will have zero product. As I keep emphasizing, scalar multiplication is the name of the operation performed by bringing them in the same direction. So, you multiply numbers on the same axes.

Vector multiplication tells you that the product of quantities on the same axis will be zero, and the product of vectors perpendicular to each other will be on the third set of axes according to the right-hand rule. In fact, the vector product is the name of a mathematical notation that emerged to complete the scalar product, which is incomplete when analyzing expressions that come from the nature of science and are formulated. Since the vector product and the scalar product complement each other, the *puzzle* pieces are completed.

When the vector multiplication rule was formed, a relationship was established between the area of the parallelogram formed by the vectors. The product of the magnitude of the vector at the base and the perpendicular side of the other vector with the sine angle gives the area of the parallelogram in between. In this way, a dependency relationship was established between the vector product and the area calculation in geometry.

It is in the spirit of mathematics and geometry to understand the area of a parallelogram as the product of vectors, and the operation is just a definition. When you organize the multiplication by considering that since vectors in the same direction will not create an area, their vector product will be equal to zero, you arrive at the multiplication rule in the matrix.

Line equations in the coordinate system can also be written for vector and scalar quantities. As you know, both vector and scalar quantities can be represented by lines in the coordinate system. Therefore, it does not matter whether the problems transferred to the coordinate system are vector or scalar. Ultimately, it is you who defines the problem and builds meaning from the products of functional relations.

Determinant is a concept that emerged with the vector product of two or more vectors. Although the determinant gains meaning with the vector product and takes its rules from there, it is not only used in solving vector problems. Because only the multiplication rule is taken from there. It gains its real identity through the coordinate system like a matrix. You can carry not only vector operations on the coordinate system, but you can also carry scalar operations there.

Finding the area of the parallelogram between the determinant and the vectors makes many problems understandable. Of course, the determinant can also be zero. In some problems, you may even need to perform an operation for the determinant to be zero. Zero means that the line segments are in the same direction. If it is negative, the concept of direction comes into play here, the area is still the same, but the direction of the vector representing the area is opposite.

I deliberately used the definition of a vector representing the field, because it is a consequence of the assumption that "the product of a vector and a vector is still a vector". You may remember from derivatives and integrals that the integral or derivative of a function is also a function. The area of one function can correspond to a linear quantity in another function. If you remember, speed is the derivative of the path, while acceleration is the integral. So, the line segment resulting from the area of the acceleration function or the derivative of the path function can represent the same thing. For example, let's say the area of a function is 15 square units, this value can be represented as a length on one of the axes of another function in the coordinate system. In other words, the magnitude we call area can be represented and represented in the coordinate system as a line segment or even as a point. What I am trying to emphasize here is the fact that the spirit of Linear Algebra is everywhere in mathematics.

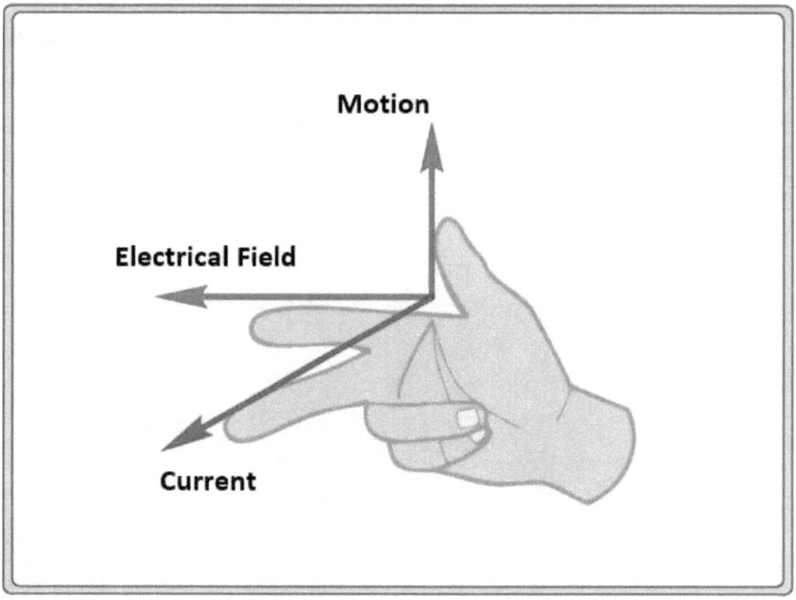

Figure 10.11: The right-hand rule.

We say that the third vector resulting from the product of two vectors gets its magnitude from the area of the parallelogram. If you ask, "How do we determine the direction of the vector resulting from the product?", you can see that the concept of direction here comes from the right-hand rule you learned in physics lessons. "A magnetic field is created where an electric current pass, and the direction of the magnetic field is determined by the right-hand rule," you may remember from physics lessons. Anyway, didn't mathematical rules arise to explain physical expressions? So, if there's a contradiction here, it's a problem, isn't it? Why does traffic flow on the right? Because it is a rule. Traffic could also flow on the left, and there are countries where it does. But we put traffic signs and other rules according to the main rule. Just like this, it is said that "The direction of the new vector formed by vectorial products of vectors is determined by the right-hand rule".

Since the two vectors meet in a plane, the new vector resulting from the product will be perpendicular to this plane and its magnitude will be equal to the area between the two vectors. This is the definition of the right-hand rule. The rule that the new vector resulting from the product of vectors is perpendicular to the other vectors is based on a rule that emerged from the physical phenomenon of electromagnetic forces.

However, no special effort is needed to create this rule. In other words, instead of the right-hand rule, the left-hand rule could have been established. It would not be right to attach too much meaning to this. Now this rule has been established and you need to base all operations on it. Otherwise you will have a lot of contradictions in problem solving. If you encounter an experimental result that does not obey the right-hand rule, for example, if you have a data set that gives an inverse direction in multiplications, don't panic! Then we just need to put a minus sign in front of the product in the formula. So, it is not correct to expect every physical phenomenon to be like electromagnetic forces. If there is no reverse direction or perpendicularity, then you may need to use trigonometric expressions like sine and/or cosine. As long as you have the experimental data, mathematics will do you the favor you need to extract meaningful expressions from it.

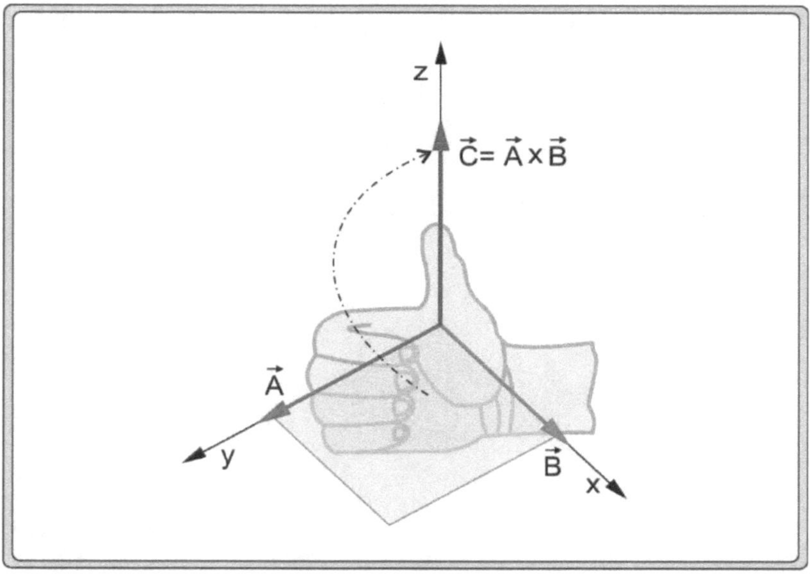

Figure 10.12: Representation of vector product with right hand rule.

The right-hand rule is a very valuable analogy. Understanding physical issues with the limbs of our body facilitates the solution and perception of problems. Why do you think finger and inch calculations are taught from primary school onwards? Science, which touches life, is actually enriched by such definitions and is very easy to learn. This does not mean that we should change our unit system. If we internalize the current unit system that we have been using for almost a hundred years, we will solve all problems again.

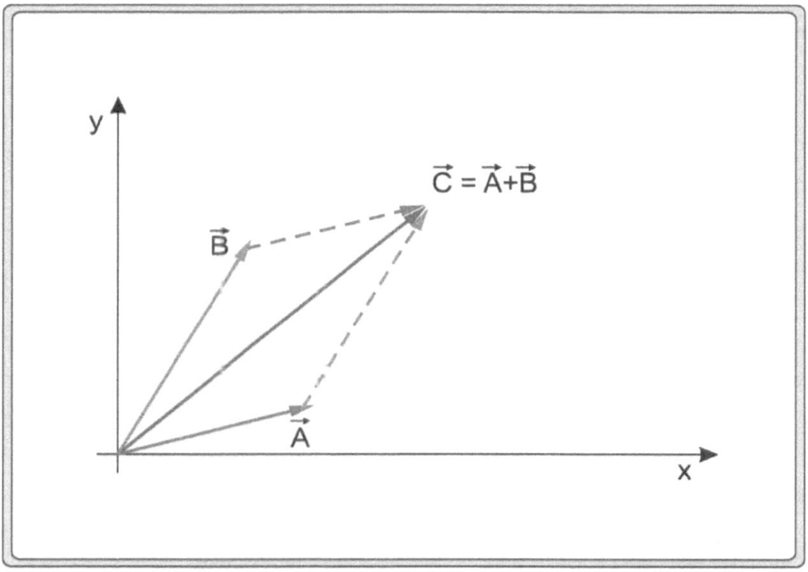

Figure 10.13: Sum of Vectors.

When solving physics problems, especially when you look at the formulas we use, you see four operations everywhere. It is almost impossible to analyze and interpret a problem that is not reduced to four operations. When four operations are combined with geometry, the problem solution gains a completely different dimension. This combination plays a very important role in the proliferation of mathematical expressions.

When you multiply a number by its inverse, you get 1. We use this rule to solve many problems. For example, if 3 apples cost 18 liras and you want to find the price of an apple, you can find the price of an apple by finding the inverse of 3 with respect to multiplication and multiplying both sides of the equation side by side. As you know, the inverse of 3 by multiplication is 1/3. Although the inverse operation is hidden because we directly divide 18 by 3, it is actually a very common operation done by mathematicians. Since the determinant represents an area, it finds its meaning in multiplication.

When you look carefully at the functional relationships, you can see multiplication in many places. The rule of inversion in a matrix is based on the rule of taking the inverse of a number by multiplication, as I have just explained. Of course, in matrix notation, which gives meaning to multiple functions, the inverse of multiplication is a very important step in the analysis of the matrix.

As I mentioned before, we use 1 for inversion according to the multiplication you do in the classical way. To find the inverse of a matrix, we use the unit matrix. To get the unit matrix, you will need the determinant. I want to explain the reason for this using the area of a rectangle. As you know, the area of a rectangle is equal to the product of the sides. Now let's place a rectangle on the coordinate axis. Let the side lengths be 4 and 7. Then the area of the rectangle will be 28. This means that the determinant of the matrix expression that gives the area of the rectangle is 28. Now let's divide these side lengths by the area to get a new rectangle. The side lengths of the new rectangle are now 1/4 and 1/7. If you write this in matrix form, its determinant, i.e. its area, is 1/28.

You can easily see that the area of the new rectangle obtained as a result of the side lengths you found by dividing each side by the area is equal to 1 time the area of the first rectangle.

There is a very simple rule behind this. Let's call the area c of a rectangle with side lengths a and b, then $c=a*b$. Now divide each of the sides by the area and you get a new rectangle with side lengths a/c and b/c and its area is $a/c*b/c=ab/c^2$. Since the area of the first rectangle was $ab=c$, if you simplify, you will find the area of the new rectangle to be $1/c$. The area of the first rectangle is c and the area of the new one is $1/c$. When you write these areas in matrix form, you can see that they are both equal to their determinants and their product of determinants is equal to 1.

What I want you to see here is the fact that the area between the new edges obtained by dividing the edges by the area is the inverse of the original area. If you see this, you immediately realize that the determinant is actually a tool for inversion with respect to multiplication. You use the area as a tool in problem solving. You can think of it as a kind of proportion constant. What I am trying to explain is that the concept of area is a concept that emerged to establish a concrete relationship between the elements of the matrix. This is what we call a determinant. Thanks to the determinant, we establish a dependency relationship between the functions that make up the matrix through its edges and its area, and we derive meaningful expressions from them.

The determinant tells you the same thing whether it is for a vector or scalar problem. Because for both, it tells you that it is an expression that gives you the area between the functional expressions linearized into the coordinate system. This area is the reality that establishes the most concrete connection between the elements that make up the matrix.

If the number lines do not intersect, that is, if they are in the same direction, of course the determinant will be zero. If the problem is vectorial, the physical meaning of a zero field is that, for example, if the electric current and the motion are in the same direction, there will be no magnetic field.

For scalar expressions, a zero field means that the problem has no solution. For example, you have two equations. If 3 apples and 2 oranges cost 5 liras in the first equation and 6 apples and 4 oranges cost 10 liras in the second equation, it tells you that there is no solution. You can multiply this problem; let's say you traveled 5 km in 3 hours at x km/h and then 2 hours at y km/h in the first equation. In the second equation, when you travel 10 km when you travel at x km/h for 6 hours and then y km/h for 4 hours, you cannot find what the x and y speeds are with these equations. Let us briefly illustrate this problem with the equations below:

$$3x + 2y = 5$$
$$6x + 4y = 10$$

As you can see, the second equation is the first equation multiplied by 2. This means that there is only one equation and therefore no solution to this problem. Let's represent these expressions in matrix form:

$$\begin{bmatrix} 3 & 2 \\ 6 & 4 \end{bmatrix}$$

The determinant of the matrix on the previous page will be zero. What does zero mean? No solution. In fact, these results tell you that your equation is incomplete. A zero determinant tells you that the equation in two unknowns has one equation when it should have two different equations. In geometric terms, it tells you that two lines are in one direction when they should be in two different directions, that is, the lines are parallel to each other. But don't be confused when I say that there is no solution here! Maybe this is the solution you are looking for. In other words, when the determinant turns out to be zero, you have found where the problem is in the system.

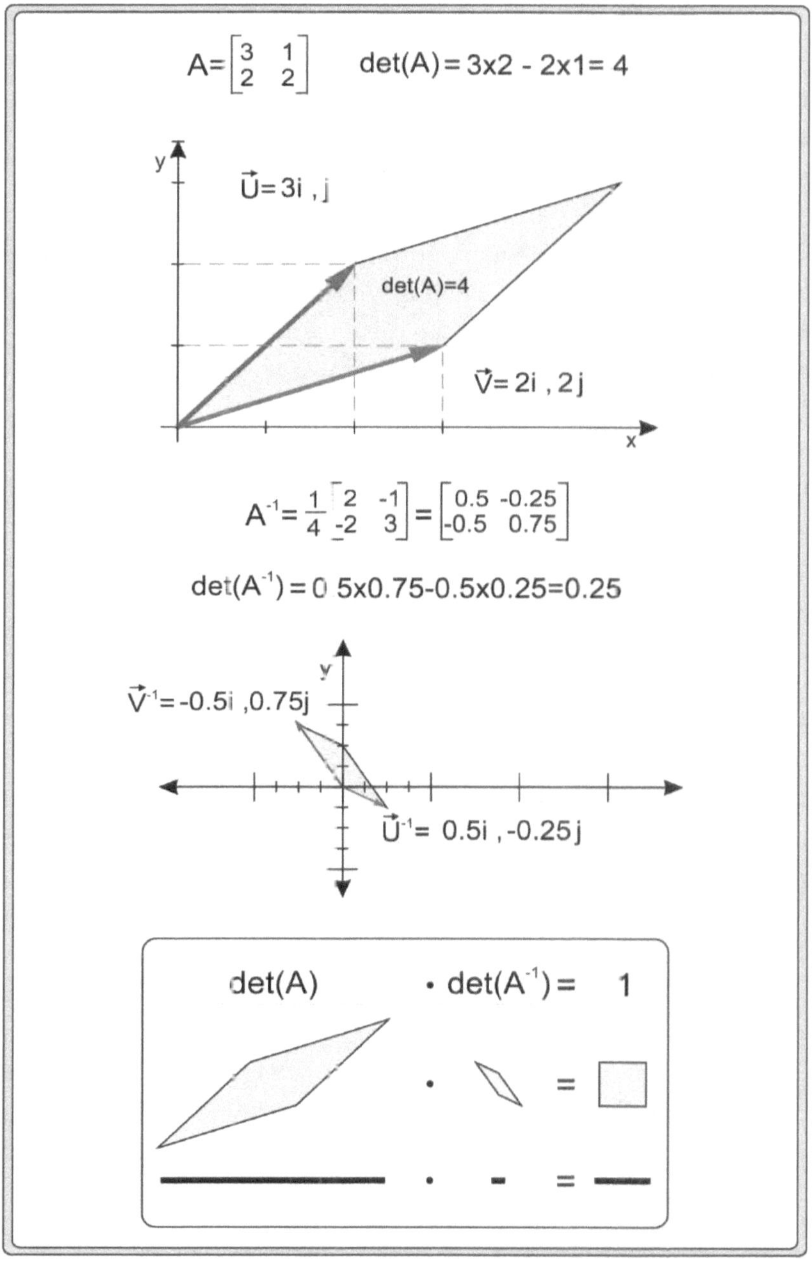

Figure 10.14: The product of the determinants of a matrix and its inverse is equal to 1.

If you read addition, subtraction, multiplication or division in matrices in terms of vectors, your job will be much easier. We can represent vectors on the coordinate system in maximum 3 dimensions. The matrix extends the boundaries of this three-dimensional space. This is the main reason why the matrix is so useful.

So, for example, you may not be able to represent a 4x4 or 10x10 matrix in a coordinate system. However, being able to make sense of multidimensional systems like this with the rules of the matrix is a wonderful gift of mathematics. In this expanding space thanks to the matrix, we can even say that there is almost no problem that cannot be solved.

Of course, the multiplication of matrices much larger than 3x3 matrices is derived from one of the most fundamental rules of mathematics, the property of multiplication to distribute over addition. Whatever the subject matter, the derivation of mathematical concepts does not go beyond the limits of the four operations.

I am aware that I have been emphasizing functions and the coordinate system since the beginning of the book. The main reason for this is the fact that being able to represent functions in a two- or three-dimensional coordinate system allows many mathematical concepts to emerge. If you don't fully grasp a mathematical rule, then move it to the coordinate system and start thinking about it. Eventually you will see that something meaningful will emerge.

When I was not aware of the deeper meaning of mathematics, I used to think that problems were solved by placing the real images of objects in the coordinate system. In other words, I would place a car, an airplane or anything similar to these in the coordinate system and think, "This is enough for me!" and I thought that mathematical concepts were derived from these shapes. It did not occur to me to make sense of the formulas and functional relationships that emerged from problems. It was only much later that I realized that the formulaic expressions of a problem and the resulting graphs were very different expressions from the actual pictures of objects.

In fact, it is best to explain the subject with a question like "What is problem solving?" Isn't it more important to understand what is a structural strength problem, a problem related to the shape change of materials with the effect of heat, or a problem where force or power is questioned? If you are going to build an airplane engine, then when you approach the problem as "How much power will it produce?" doesn't the engine already start to take shape from here? With the question "How much power do you need?", your geometry starts to take shape and the dimensions of your fan diameter start to emerge accordingly. Your engine is shaped by these functional relationships. By transferring it to the coordinate system, you start to analyze and solve your problem by detailing these shapes. If you use math well, you will eventually reach the engine.

Of course, you can see problems in physical formulas and then mathematical expressions. The meeting of the scalar and vector quantities of formulas with the four operations has given us many concepts. When you see functions in derived quantities and matrix structures that emerge with their multiplications, you realize once again that mathematical expressions actually emerge to make sense of real problems.

If you had asked me to summarize in one sentence what I have been telling you since the beginning of the book, I would have said: "Just look at the formulas, they tell you everything." Try to understand the expressions there, then observe how they meet the four operations, try to see the relationships in their continuous changes. $f = ma$'ya,
$e = mc^2$ or F=$G\frac{m_1 * m_2}{r^2}$ Look at formulas like this. Try to understand them by asking questions like "Why are we multiplying, why are we dividing, why are we squaring?". Then you will understand better what I mean.

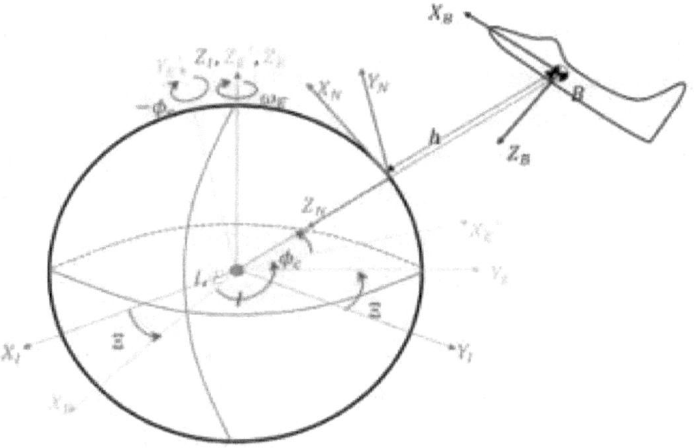

u, v, w ⟶ **Axial velocity vector**

p_x, p_y, p_z ⟶ **Position vector of the aircraft**

A=

u	v	w	p_x	p_y	p_z
-0.7356	-49.7137	45.2815	-87.1740	0	3.951e-4
-0.1318	-0.0981	-0.8476	0	0	4.2663e-6
0	-0.6678	-0.4985	0	0	0
0	0.0000	1.0000	0	0	0
0	0	0	0	-1.0000	0
0	-172.0000	0	184.0000	0	0

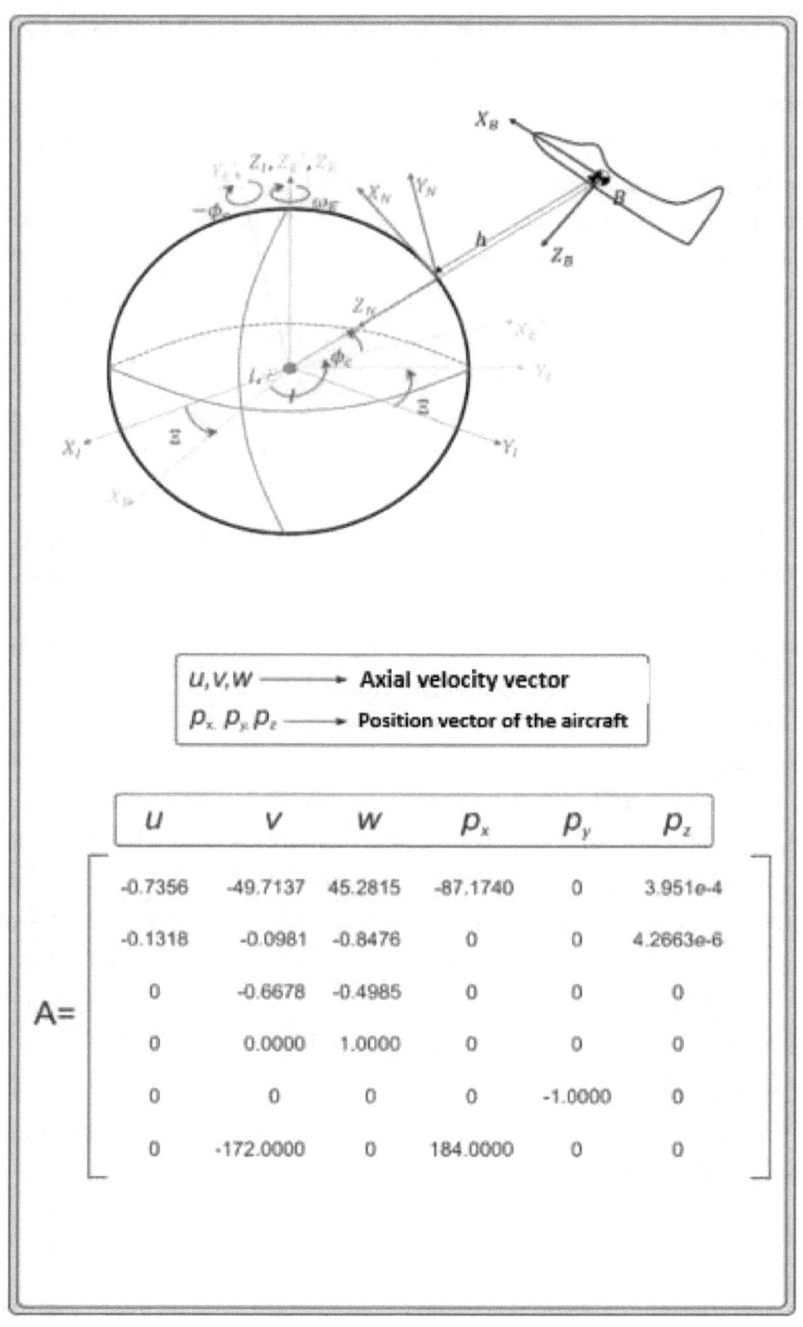

Figure 10.15: An example flight control matrix representing the level flight of a UAV.

As an example, I want you to look at the flight control matrix of a UAV above. Notice that the velocity and position vector are processed simultaneously thanks to the matrix. If you have a control problem, I want you to see that the control is decrypted by processing two different states such as speed and position at the same time. Thanks to the matrix, you get the opportunity to find solutions where different parameters are processed at the same time and these parameters are not mixed together.

I mentioned that the arrow in vectorial notation is an expression that you should pay attention to when doing four operations. Although the result of vector and scalar multiplication says different things, since a functional relationship, whether vector or scalar, turns into meaningful expressions in the coordinate system, once the expressions taken from here enter the matrix table, it loses its meaning whether it is scalar or vector.

Let's try to understand how the matrix is used by showing a simple problem about scalar quantities in a matrix. For example, "Let 2 loaves of bread and 3 eggs be 13 cents; 3 loaves of bread and 4 eggs be 19 cents. How many cents is one bread or one egg?" Let's make a small matrix application through a question. We have been setting up such equations since middle school. We can define this problem with the following expressions:

$$2x + 3y = 13$$
$$3x + 4y = 19$$

Now let's solve this problem with the classical approach. For this, if we expand the first equation by 3 and the second equation by 2;

$$\begin{array}{rr} & 6x + 9y = 39 \\ - & 6x + 8y = 38 \end{array}$$

If we subtract them side by side, we find y to be 1. If we substitute y into one of the two equations, then you can easily find that x is 5.

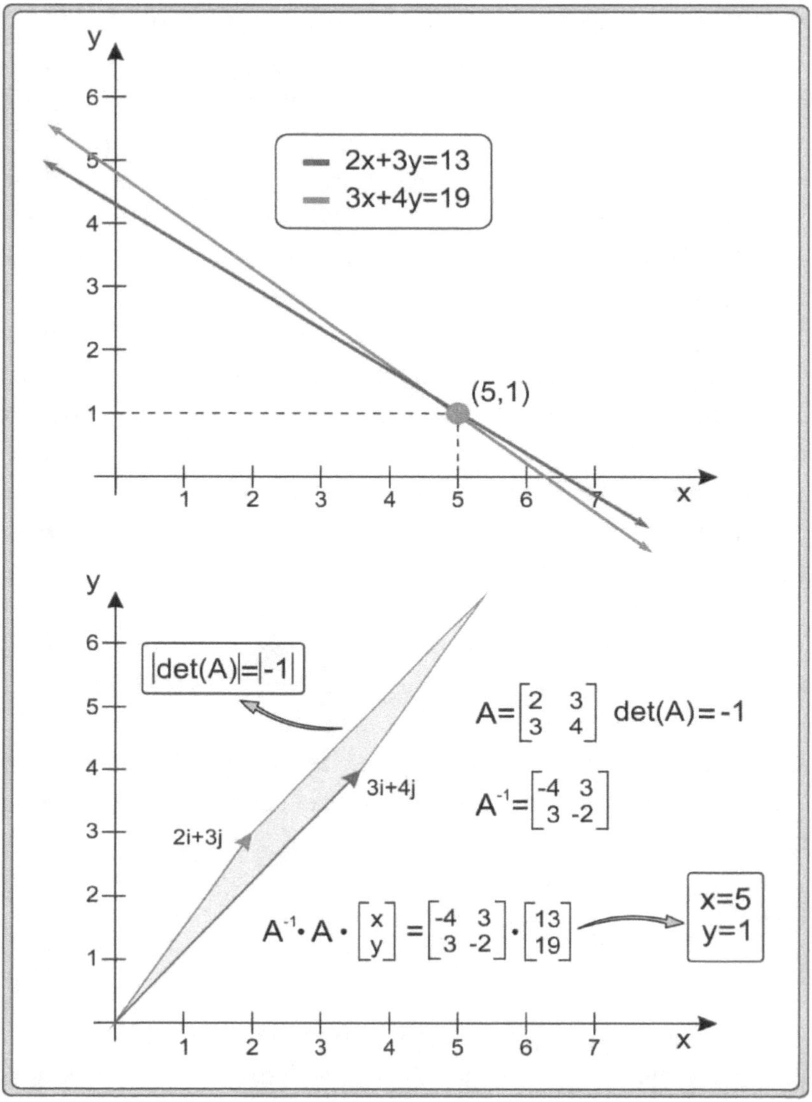

Figure 10.16: Solution of a simple problem using a matrix.

The key point in solving engineering problems is to convert them into expressions that can be transferred to the coordinate system. If you transfer the quantities to the coordinate system, the math will take care of the rest.

Although the determinant takes its soul from a vector operation, it takes its real identity from the coordinate system. Because you deal with the shapes of the expressions you transfer to the coordinate system. Here, you try to get meaningful expressions with the edge, intersection, area or volume of geometric shapes. The determinant helps you to get meaningful expressions here.

If you go back to high school, you are taught that the equation of two lines passing through a point in a coordinate system is given by the coordinates of that point. Of course, this is a very correct and basic approach. When you transfer the equation of two lines to a coordinate system, it is in the spirit of mathematics that where two-line segments intersect is the solution set. With this most basic rule, you can find solutions to problems.

$$2x + 3y = 13$$

$$3x + 4y = 19$$

The reason why the solution of the above two equations exists is because its determinant is not zero. If the determinant were zero, it would mean that these two equations are in the same direction. There is no concept of area from two-line segments in the same direction. So, the product of two vectors in the same direction is equal to zero. But when you take the determinant here, you can see that it is -1. This means that the equations are not in the same direction, the lines will intersect somewhere in the coordinate system. This intersection is already your solution set.

The fact that the equations are not in the same direction means that you can take their inverse when you put them in matrix form. Because if you are looking for a solution in matrix form, you first take the inverse of the matrix according to multiplication. Just as you cannot get the result when you divide a number by zero, a matrix whose determinant is zero does not have an inverse.

Now you know that matrix notation is actually a neat representation of numbers in a table without any confusion. If there were no matrices, it would be very difficult to get out of multivariable expressions with many unknowns. We would be confused and we would not be able to see what is where.

For Linear Algebra, which gives the soul to mathematics, you can say that it is the art of making sense of relationships complicated by curves with lines. Line segments create a much clearer picture in our minds and you can connect line segments much more easily. If you look carefully at this world, you will see that most of what we do is to find area and volume. Now forget about the units, just combine these lines with lengths and you will understand what I mean.

Linear Algebra has lines in its soul. Mathematics reveals all the relationships behind these lines. Mathematics is not so much interested in their units. When units come into play, you enter the world of science and social sciences and according to the definitions there, expressions find their true meaning. If you are only a mathematician, you cannot fully understand what life wants to tell you. That is why you need to enter the world of science and social sciences with mathematics. Units give value and meaning to all relationships in this world. If you know these units well, you will immediately understand what they mean through mathematics.

I would like to explain the *eigenvalue* and eigenvector, which are the most valuable subjects of the matrix. Without the matrix, we would have difficulty in performing operations such as multiplication and division between very large groups of numbers and in extracting meaningful relationships from them. With the introduction and development of the computer, the value of the matrix has been better understood and started to be used much more effectively.

In the spirit of Linear Algebra, I said that if we can transfer a problem to the world of lines, we can make it more understandable there. How did we make functions understandable while explaining derivatives and integrals? Of course, when we break functional relationships into line segments, it was very easy to add, subtract, divide and multiply them.

Regardless of the physical meaning of your problem in systems of equations defined by scalar or vector quantities, eigenvalue and eigenvector appear as a concept that makes the relations between equations meaningful in solving a problem and enables a proportional connection between them. In the literature, eigenvalue is defined as the expression that maps a matrix to a scale. In this section, I wanted to make the eigenvalue and eigenvector more understandable, since I see that we cannot get anywhere with this memorization in the literature.

It is best to perceive this by emphasizing that each of the rows of a matrix corresponds to a function. Bringing these functions in the same direction actually means finding out if there is a solution to a physical problem. For example, if you are solving a magnetic field problem and the electric current and the motion are in the same direction, you know that there is no magnetic field. You know this from the fact that if the vectors are in the same direction, their product is zero.

We also transform scalar quantities into meaningful expressions with lines. Therefore, it is not correct to look for eigenvalues and eigenvectors only in vector quantities. Therefore, you will encounter the same result in scalar problems. Since no two lines oriented in the same direction will intersect in space, you will never find a point in these equations that will give the same line equation.

It is a very important step in problem solving to see that expressions brought in the same direction have no solutions, or in other words, that they have infinite solutions. An infinite solution means that there are not two equations but one equation and you can see that every solution will satisfy the equation. If 2 apples and 1 orange cost 10 cents and 4 apples and 2 oranges cost 20 cents, we are talking about an equation. Then we can produce infinite solutions such as if the apple is 1 cents, the orange is 8 cents, if the apple is 5 cents, the orange is 0 lira, if the apple is 3 cents, the orange is 4 cents.

Eigenvalues and their eigenvectors are expressions that bring the functions into the same plane and direction. If you ask "What is in these results?", they actually tell you whether there is a proportional relationship or a connection between the functional relations that make up the matrix.

To find the solution to a problem you need to be in the same world, and that's what eigenvalues allow you to do. The reason why I say the same plane is due to the fact that the eigenvalue can also be a complex number. Otherwise I would say the same direction. As you know, complex numbers are very important mathematical expressions that move the number line to a plane. Complex numbers are used a lot in physics problems. As the problems get deeper, you will see that the number line is not enough. You will understand even better how complex numbers come to the rescue by moving a number line to a plane.

In vibration problems, which is one of the most important topics of engineering problems, it is very important to find the place where the system resonates and you find the resonant frequency with eigenvalue problems. The expression resonance, which means that the amplitude goes to infinity, confirms this definition. Because for vibration, resonance is the name of the place where the amplitude around a frequency always goes to infinity. Depending on the type of problem, eigenvalue problems often allow you to find regions where there are infinite solutions. Imagine you find a frequency value and this value always satisfies the equation. This is a dangerous situation and you find that you have to design the system away from that frequency value.

If you remember from limits, an infinite limit means no solution. Remember also from series, we always want the series to be convergent. If the series is divergent, it's useless. So you cannot check an expression that goes to infinity. Look at eigenvalues and eigenvectors from this point of view! You need to find the places where the functions converge to the same direction and plane. Remember from electrical problems, if you find where there is a short circuit, your circuit works fine. In the same way, when you find and repair short circuits in problems, you often solve the problem.

If we look at the mathematical definition of eigenvalue and eigenvector; A being a matrix, eigenvalue and eigenvector being a unit matrix, we call the scalar number X that satisfies the equation Ax= Xx an eigenvalue. Now, when solved as (A-X)x=Ix, you can reach X, the eigenvalue. When you multiply the X's you find by x, you find the eigenvectors. Why do you do all this in a matrix? Because you move two or more relations in a system of equations in the same direction and plane. I would like to illustrate what I mean with a small example.

$A = \begin{bmatrix} 4 & 5 \\ -7 & -8 \end{bmatrix}$ and eigenvectors $x = \begin{bmatrix} x_1 \\ x_2 \end{bmatrix}$ Get it.

Now $(A - \lambda)x = Ix$ Let's find the eigenvalues in the expression.

$\begin{bmatrix} 4-\lambda & 5 \\ -7 & -8-\lambda \end{bmatrix}$ when you take the determinant of this matrix

$(4-\lambda)(-8-\lambda) + 7 * 5 = 0$ you get.

If you edit this statement,

$\lambda^2 + 4\lambda + 3 = 0$ is obtained.

This equation has two roots, and when you solve it λ_1 =-3 and λ_2 =We find -1. Now when you put these expressions into the matrix, you get two vectorial expressions in the same direction.

The first one $(A - \lambda_1) = \begin{bmatrix} 7 & 5 \\ -7 & -5 \end{bmatrix}$,

and the other one $(A - \lambda_2) = \begin{bmatrix} 5 & 5 \\ -7 & -7 \end{bmatrix}$.

As you can see, take a look at the rows of the matrix and you will see a proportional relationship. In the first matrix you can see that the ratio of each column to each other is equal to -1 and in the second matrix it is equal to -5/7.

What I also want you to see here is that we have obtained 2 functions with different coefficients. Notice that thanks to the eigenvalue, these turn into 2 functions in the same direction. As you can see from the fact that the determinant is zero, the eigenvalue transforms the two functions into expressions in the same direction.
When you take the eigenvalues into functions and transfer them to the coordinate system, you get quantities on the same number line but in opposite directions. We call these quantities eigenvectors. Let's take a look at how to find eigenvectors using eigenvalues;

$Ax = \lambda x$ eigenvalue from the expression $\lambda_1 = -3$ and take it;

$\begin{bmatrix} 4 & 5 \\ -7 & -8 \end{bmatrix} \begin{bmatrix} x_1 \\ x_2 \end{bmatrix} = -3 \begin{bmatrix} x_1 \\ x_2 \end{bmatrix}$ regulate equality,

$4x_1 + 5x_2$=-3x_1

-7$x_1 - 8x_2$=-3x_2 you'll find it.

When you edit these expressions $-7x_1 = 5x_2$ you get.

$$\frac{x_1}{x_2} = \frac{-5}{7}$$

Eigenvalue $\lambda_2 = -1$ if you take

$\begin{bmatrix} 4 & 5 \\ -7 & -8 \end{bmatrix} \begin{bmatrix} x_1 \\ x_2 \end{bmatrix} = -1 \begin{bmatrix} x_1 \\ x_2 \end{bmatrix}$ From Equality

$4x_1 + 5x_2$=-x_1

-7$x_1 - 8x_2$=-x_2

When you edit these expressions $5x_1 = -5x_2$ you get the expression.

$5x_1 = -5x_2$ from the expression $\frac{x_1}{x_2} = \frac{-1}{1}$

Namely $x_1 = 1$ If it is $x_2 = -1'$dir. If we show this as a vector:

$$x_1 = 1 \text{ and } x_2 = -1$$

$\frac{x_1}{x_2} = \frac{-5}{7}$ And the expression;

$x_1 = 1$ If it is $x_2 = \frac{-7}{5}$ becomes. If you do this vectorially,

$$x_1 = 1 \text{ and } x_2 = -7/5$$

in the form of a line. But remember that the lines are in the same direction!

Eigenvalue and eigenvector tell a proportional relationship between variables in the same direction. Being negative means that the quantities are only in opposite directions.

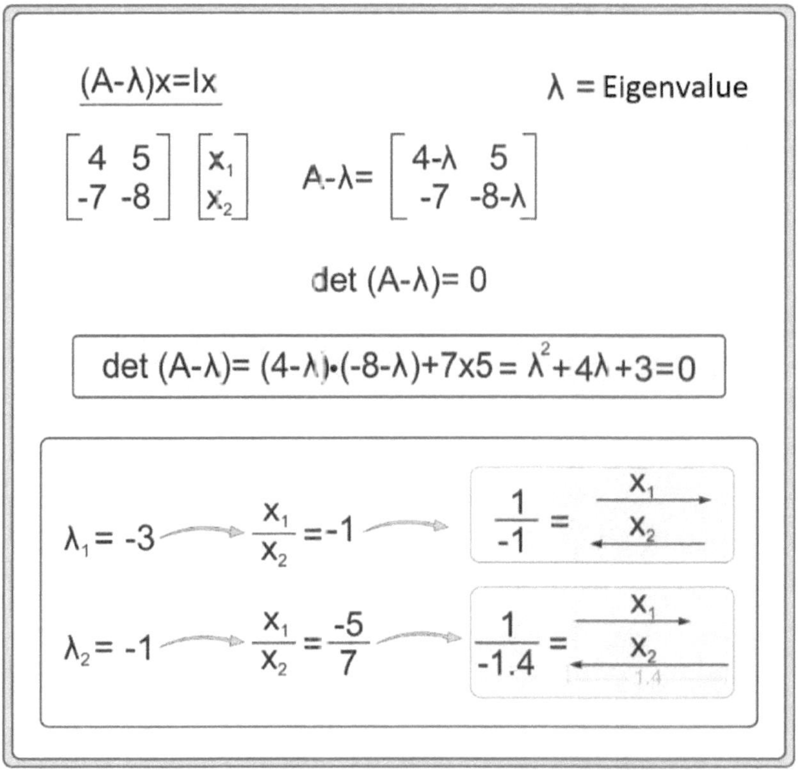

Figure 10.17: A simple example for eigenvalues and eigenvectors.

Let us conclude by emphasizing that a matrix is a table of numbers that allows you to establish meaningful relationships between numbers in a coordinate system. We enter the world of problems through functions. With functions, two basic quantities, scalar and vector, enter our lives. Whether vectorial or scalar, transferring these quantities to the coordinate system and obtaining expressions in matrix form with them is the most important achievement in creating a mathematical model of a real problem.

I tried to explain that if there is more than one functional relationship, then the matrix offers very effective solutions. Whether it is the determinant or eigenvalues and eigenvectors, I think you now have a better understanding of why these matrix tools are used to solve a problem. Of course, you may not be able to solve a very complex engineering problem immediately just by learning the background. But if you know why mathematical concepts arise and where they should be used, then solving problems will not be as difficult as it used to be.

On Conclusion

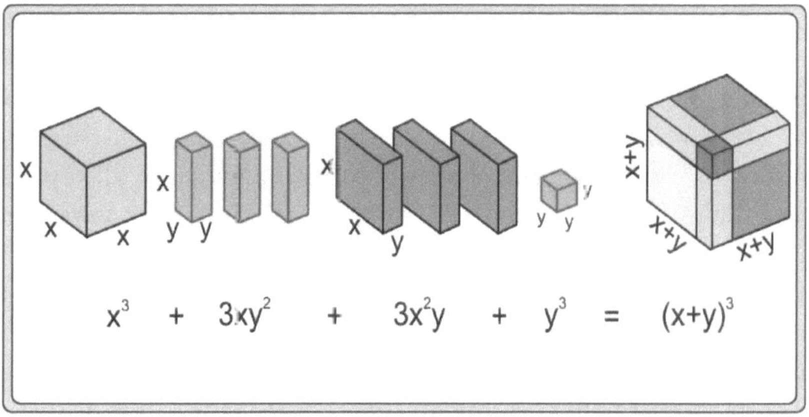

On the Outcome

The importance of mathematics education is explained everywhere without connecting it to life. Without connecting it to real life, everyone says, "Learn mathematics, mathematics opens your mind! If you want to be a doctor or an engineer, you need to know math!" Mathematics is not learned only to have a profession. It is learned to understand, interpret and improve this life we live. I hope this has met your expectations. The place of nations that internalize the analytical thinking system in the world is obvious, unfortunately, as a nation that has difficulty in connecting with this life, we need to accept our place in the world and act to change it.

While writing this book, I had the opportunity to examine the math questions that my daughter, who was preparing for high school, was studying. I saw that puzzle questions were being asked under the name of "new generation questions". I read a question three times and still could hardly understand it. Mathematics is based on understanding what you read. Without understanding the background of mathematics and connecting it with the world you live in, the education given is not very useful. When I examined the questions in depth, unfortunately, I saw that they were again based on memorization.

When asked "Why do you learn exponents, radicals, ratios and proportions?", if you don't have an educational structure that explains that all these are taught in order to understand and interpret basic mathematical concepts such as derivative, integral and matrix, which I explain in this book, then you don't have the spirit of mathematics. If you cannot see the derivative and integral through proportion, or if you cannot make the transition from exponential roots to logarithms, then the deeper meaning of mathematics has not appealed to your world. It really upset me that a question asked years ago, "Why do we learn these things?" still cannot be answered properly. You cannot come to the real world through the questions. By asking such questions, we can never give the deep meaning of mathematics to our children. If people cannot say "this is why I learn mathematics", then there is memorization. Let me repeat it again: The problem is not in the structure of the curriculum, it is in its spirit. Mathematical education without that spirit will never achieve its goal.

As I mentioned while explaining the matrix, sometimes you convert the coefficients of the line equation into points in the coordinate system and look for the solution of a problem here. While explaining these coefficients, I remembered the expression $\Delta=b^2 -4ac$ that we use for quadratic equations formulated as $ax^2 +bx+c=0$, which I solved a lot in high school. I can say that finding the roots of the equation using this expression was my biggest hobby in my high school years. Whenever I came across a quadratic equation, I would immediately look at the coefficients and try to solve the problem. I never thought of making sense of the relationship between these coefficients and x. I can even say that I pretended x did not exist. Isn't it strange that you can't see x in order to find x? Why don't we immediately think of relating these coefficients to x? Actually, Δ is not very important, it is a tool. The important thing is to find x. Wouldn't it be a nice nostalgia for you if I told you that after multiplying each term in the equation by 4a, adding b^2 to both sides of the equation

$ax^2 + bx + c = 0$	*Multiply both sides by 4a,*
$4a\,x^{22} + 4abx + 4ac = 0$	*Add b^2 on both sides*
$4a\,x + 4abx + 4ac + b^{222} = b^2$	*Pass 4ac to the other side*
$4a\,x + 4abx + b^{222} = b^2 - 4ac$	*Then edit this*
$(2ax+b)^2 = b - 4ac^2$	*Take the square root*

$$2ax+b=\sqrt{b^2-4ac}$$

$$x = \frac{-b \pm \sqrt{b^2-4ac}}{2a}$$

and obtaining the exact square in the equation, x is obtained by leaving it alone and becomes an expression of the coefficients a, b, c? In fact, if we always learned mathematics in this way, wouldn't we (wouldn't we) understand better where we need it and how to use it? When you see the coefficients, don't forget the world behind them. Otherwise, you will continue memorizing from where you memorized.

The rules we have memorized have created this perspective in all of us. Isn't this similar to looking for a missing child without looking at the picture, even though you have a picture of the child? Can you search for a child without looking at the picture? Just as you cannot search without knowing what you will find, you cannot teach mathematics without knowing why it is taught.

In analytical geometry classes, the terms "multidimensional space" or "spaces" are used a lot. After reading this book, I hope you will no longer confuse the space in the mathematical world with the space you live in the physical world! And what do you mean by spaces, not spaces? We can't cope with one space, where did spaces come from? I mentioned this when I was explaining the matrix. We can define the space we live in with 3 independent variables x, y, z in a 3D coordinate system.

In the mathematical world, each problem creates its own space with its own variables. This is where the multidimensional space appears, right? In fact, the spirit of independent variables travels in the plural concept we call multidimensional space. So, a problem with 4 or 5 variables means 4 or 5-dimensional space. Why does the concept of time come to the fore for the fourth dimension? Because time is one of the most valuable independent variables. We use the time variable in almost many problems. From speed problems to interest rate problems, whatever problems there are, they are all time-dependent problems. Therefore, what will be the fourth dimension if not time? So, there is nothing metaphysical behind the definition of time as the fourth dimension. Time is considered as the fourth dimension purely because of this approach.

Every problem is born, lives and dies in its own space with its independent variables and becomes meaningless in another space. I think that when you think of multidimensional space or spaces, you don't think of real space, which is a very dark place. As you know, we use the matrix to get out of multidimensional space. If you don't see this, you don't understand the importance of the matrix. Then how are we going to solve real and complex problems, right?

The PISA results actually shout the whole truth in our faces. Unfortunately, we keep looking for solutions to change this with the same methods. If we keep making structural reforms in education and yet our PISA ranking does not change, we need to stop and think seriously. Without mathematics, there is no physics, chemistry, biology or social sciences. Well, if this is the truth, shouldn't we build an education model that gives proper answers to the question "Why do we learn these things?"?

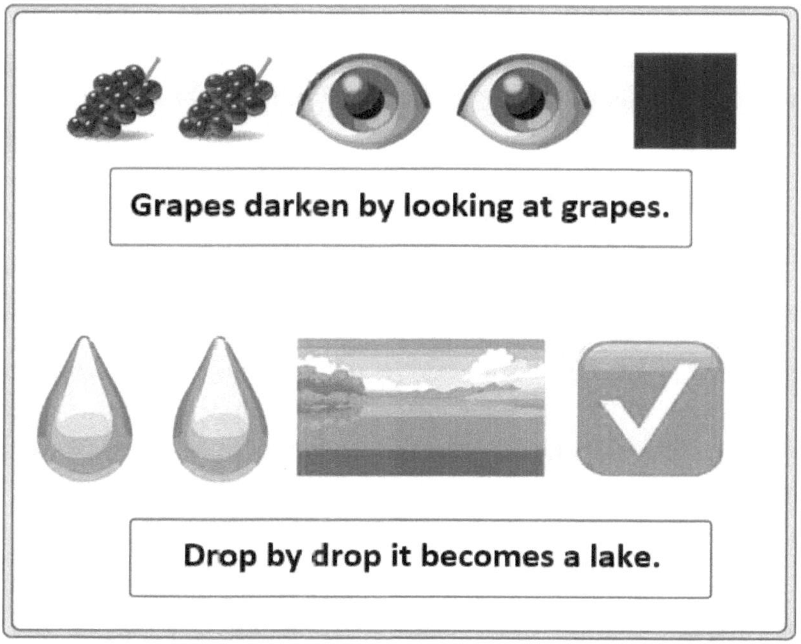
Grapes darken by looking at grapes.
Drop by drop it becomes a lake.

11.1: As long as the math is correct. The point can also be explained with just pictures.

I recently saw the visuals of a movie told with emojis. When I saw the whole picture, I said, "Here is the power of functions!" Isn't it a creative idea to use emojis representing human behavior in every scene? Just like this, we have to make mathematics everywhere in life. We have to show that mathematics comes out of everything. If you internalize what I have explained in this book, then you will understand that education has a goal and that it will carry us to a world where all these mathematical concepts are used consciously and real problems are solved.

We need an education model in which all the concepts of mathematics are clearly seen, such as "When we look at real problems, how should we establish functional relationships, what should be the type of functions, where to take integrals and derivatives, how many rows and columns should the matrix table consist of?". We need to stop teaching mathematics with the same pattern of questions and answers.

You have to know the truth behind learning the problem questions that we enjoy solving from primary school onwards. Otherwise, we will never get to where mathematics wants to take us. For example, if you are asked the question "Why are interest problems explained?" and you cannot answer this question properly, then we have a very serious problem. Now that you have read this book, you know that interest problems are not taught to calculate interest.

In simple interest, ratio and proportion and the four operations of addition, subtraction, multiplication and division open the door to the function. In compound interest, a door opens up to the number e. The spirit of the number e is multiplication together and you can understand the spirit of the number e with the most accurate compound interest problems.

When we use logarithms and choose the base according to our needs, then logarithms touch our lives. When we take derivatives and integrals in a real problem without anyone telling us to do so, then these things touch our lives. If you establish this foundation, then you will raise a generation that cannot solve problems with mathematics that touches life.

When we make sense of the wave motion of sound, we can visualize the wave due to the movement of the tongue, but when we say image, we have trouble even describing it with a photograph. Have you ever thought about the reason for this?

A "photograph" is actually the name of a fixed state. We call its accelerated state an "image". So, how can we perceive the fact that photography is wave motion? That's the real problem! Of course, you cannot look at a photograph and see movement. It is the movement of light that makes the photograph visible! What you see as a photograph is nothing but a snapshot presented to you by the light at that moment! As long as the angle of the light doesn't change, you keep seeing the same picture. Even though what you have in your hand is the same, you have witnessed that when the angle of the light changes slightly, the picture you see also changes. If the problem in perception is waves and motion, think of the speed of light. You will look for the wave motion in the photograph in the light.

In the aircraft design course, instead of analyzing the motion of the airplane in the air, the motion of the air over the fixed plane is analyzed because it gives a more beautiful and simple solution. Airplane geometry is made sense of air movement. In the same way, don't forget the movement of light, because if you stop it, you will find it difficult to see and, of course, to solve problems!

Figure 11.2: The sun takes our picture every day.

When we look at the trend of scientific development, we can say that the recording, processing and sending of sound and image to distant places has come to the forefront. You can see that image and sound are analyzed and synthesized behind all systems such as telegraph, telephone, television, satellite communication, remote sensing and night vision. The image, or picture in its instant form, was recorded before sound. The image is related to light and if there is light, there is an image. If there is no light, there is no image. When I thought about why the image was recorded before the sound, I thought of my days when my shadow appeared and I was burning in the sun. As you know, when you go out in the sun wearing a tank top, after 5 minutes it's like having a photo of the tank top taken on your body, right? So, these sunburns reminded me better that it is normal for the image to be recorded before the sound. Actually, the opposite makes more sense, but unfortunately this is the reality.

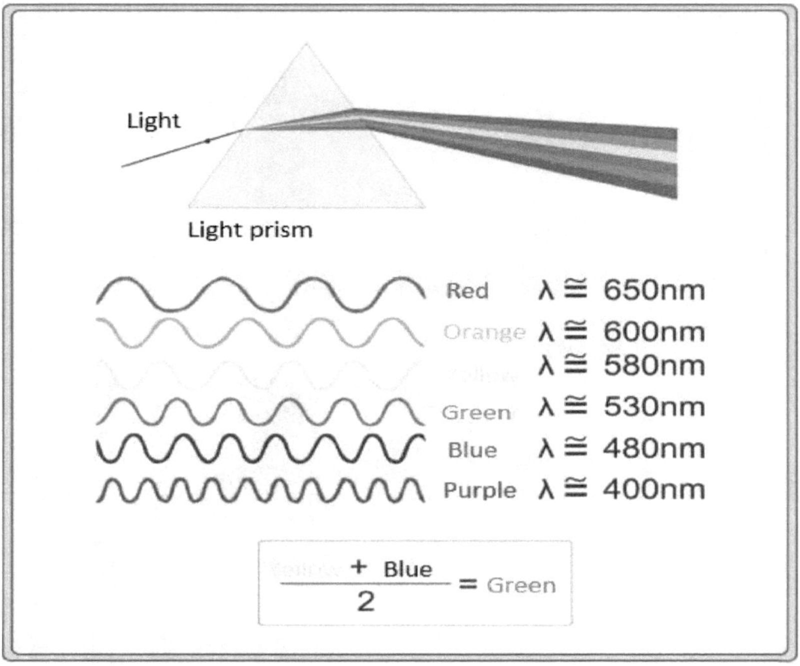

Figure 11.3: Look at addition and division in this way.

What would you say if I told you that the common characteristic of both is wave motion and that this is made understandable by the sine function? The moment electricity comes into play in sending both of them far away, we see wave motion here too, and that's why the sine function is so important.

The basic problem is not being able to see mathematics in painting, music, sculpture, photography, images. If you don't understand and interpret them in the flavor of the four operations, what you learn is no more useful than memorizing logarithms.

Now look at the wavelengths of white light in different colors in the table above; the wavelength of blue light is 480 nm and the wavelength of yellow light is 580 nm. Now add them together and divide by two and you get the wavelength of green light. Now take a pencil in your hand, color a white piece of paper blue and then add yellow to it, what color does the paper turn into? Isn't it green? If you see the summation here, you see the big picture. If you understand things in this way, then neither the addition nor the division of wavelengths will seem strange to you. When you approach real problems in this way, there are very few problems you cannot solve.

I sent this book to a friend for a comment and he asked me a question "Why didn't you add differential equations to the book?". My friend actually put his finger on a very important issue. However, since it is a dense book even in this form, I did not explain differential equations in this book, which is a book subject on its own. I have a dream of writing my own engineering mathematics book. Since I will already talk about differential equations there, I did not want to include this topic in this book.

I thought I was explaining the most basic concepts of mathematics in this book. "Differential equations are the gateway to real life," you might say. But if you say, "What's in differential equations?" you will see that there are a lot of differentiation and integration functions. The world of differential equations is a world of problems defined by expressions that include both derivatives and integrals. In fact, someone who sees the background of derivative and integral is not afraid of solving problems related to them. If you know that they are both functions, then you are not afraid to add, subtract, multiply and divide them.

Of course, I did not write this book as a test book in which mathematical problems are solved. I tried to explain where mathematical concepts come from, how they should be interpreted and where they should be used. I tried to bring explanations with my own point of view and thoughts to the questions of what thought is behind the most basic concepts of mathematics that people learn and why they learn them.

When you make sense of the formulas used in the sciences and social sciences, which have become richer over time, from the perspective I have described in this book, you will better understand how your world has become richer. You may even say, "There are so many functional relationships in life, I see functions everywhere."

If you can make mathematics a part of your daily life and connect events with mathematics, then your life will be very enriched. Stop defining and solving problems with first order functions in one unknown! It is time to use analytical thinking to uncover the truths in this life. If you know math well, you will not be afraid to deal with real scientific issues. No matter how complex your functional relationships become, no matter how high the degree of your functions, you understand that mathematics will do all kinds of good things for you to solve them.

In the mathematical journey that begins with the four operations that enter our lives from primary school onwards, I have been explaining that all the concepts produced in the world of mathematics, from ratio and proportion to radicals with exponents to derivatives and integrals, have a purpose and that when you create the mathematical model of any problem properly, mathematics will take care of everything else. In the future, I hope to meet you in a book that explains how to solve real problems with these concepts, such as how sound moves, how to record and send images, how to build airplanes, and how to mathematically model and solve problems that touch life...

www.ingramcontent.com/pod-product-compliance
Lightning Source LLC
Chambersburg PA
CBHW051343280526
45784CB00007B/2795
* 9 7 9 8 8 8 4 5 9 7 1 7 4 *